# Self Sensing Techniques for Piezoelectric Vibration Applications

Dissertation
zur Erlangung des Grades
des Doktors der Ingenieurwissenschaften
der Naturwissenschaftlich-Technischen Fakultät II
- Physik und Mechatronik -
der Universität des Saarlandes

von

Emanuele Grasso

Saarbrücken

2013

Tag des Kolloquiums:  24. September 2013

Dekan:  Prof. Dr. rer. nat. Helmut Seidel

Gutachter:  Prof. Dr.-Ing. habil. Hartmut Janocha
Prof. Dr.-Ing. habil. Stefan Seelecke
Prof. Dr. David Naso

Bibliografische Information der Deutschen Nationalbibliothek

Die Deutsche Nationalbibliothek verzeichnet diese Publikation in der
Deutschen Nationalbibliografie; detaillierte bibliografische Daten sind
im Internet über http://dnb.d-nb.de abrufbar.

ISBN 978-3-8325-3623-7

Logos Verlag Berlin GmbH
Comeniushof, Gubener Str. 47,
10243 Berlin
Tel.: +49 (0)30 42 85 10 90
Fax: +49 (0)30 42 85 10 92
INTERNET: http://www.logos-verlag.de

# Acknowledgements

The presented work has been developed during my occupation as a researcher at the Laboratory of Process Automation (LPA) of the Saarland University in Saarbrücken.

My special gratitude goes to Prof. Dr.–Ing. habil. Hartmut Janocha, chief of LPA, who has supported this work and helpfully led me through it up to its completion.

Many thanks also to Prof. Dr.–Ing. habil. Stefan Seelecke for his important advices for improving my work, and to Prof. Dr. David Naso of the Polytechnic of Bari, Italy, for having cared and shared my research results.

I would like also to thank all my collegues at the LPA department: Benedikt Holz, Chris May, Thomas Würtz, Jürgen Klesen and Leonardo Riccardi. Each one of them has contributed to create a nice and unified working environment, precondition of all successful works. I would also like to mention and thank the students who have helped me to reach this goal, in particular Ilaria Girolami and Nicola Totaro.

I dedicate this thesis work to my sister Isabella and to her two daughters, Irene and Noemi.

# Abstract

Self sensing techniques allow using a piezoelectric transducer simultaneously as an actuator and as a sensor. Such techniques, based on the knowledge of the transducer behavior and on the measurement of its electrical quantities, i.e. voltage and charge, are able to reconstruct its mechanical sensory information. In vibration control applications piezoelectric self sensing actuators are highly desirable as they allow precise collocated control. Past research work was mainly based on the linear behavior of piezoelectric materials, thus restricting the operating driving voltages to low values. This thesis addresses the problem of using a self sensing piezoelectric actuator at its full driving voltage range. A new self sensing technique is proposed, which is based on the hysteretic modeling and identification of the piezoelectric transducer capacitance. After providing a sound presentation on piezoelectricity and vibrating structures, the most common self sensing techniques are discussed and the new self sensing technique is introduced and theoretically compared to typical linear methods. Finally, after a short survey on the electronics needed for the real implementation of such techniques, the new self sensing technique is experimentally tested and compared to a typical linear one, both on a cantilevered beam and on a plate structure. The experiments have proven that the new technique is able to have good performance at low as well as at high driving voltages, differently from the linear one.

# Kurzzusammenfassung

Self sensing Verfahren ermöglichen es, einen piezoelektrischen Wandler gleichzeitig als Aktor und als Sensor zu benutzen. Diese Verfahren, die auf der Kenntnis des Wandlerverhaltens und auf der Messung seiner elektrischen Eigenschaften (bspw. Spannung und Ladung) basieren, können die mechanischen Sensorinformationen rekonstruieren. In Anwendungen der Schwingungsregelung sind piezoelektrische self sensing Aktoren sehr wünschenswert, da sie eine präzise kollozierte Regelung ermöglichen. Bisherige Forschungsarbeiten beschäftigten sich hauptsächlich mit dem linearen Verhalten piezoelektrischer Materialien, was die Betriebsspannung auf geringe Werte beschränkt. Da hierdurch die aktorischen Fähigkeiten begrenzt werden, befasst sich diese Arbeit mit der Aufgabe, den piezoelektrischen self sensing Aktor in seinem gesamten Betriebsspannungsbereich zu nutzen. Eine neue Methode, welche auf der Modellierung der Hysterese und der Identifikation der Kapazität des piezoelektrischen Wandlers basiert, wird vorgestellt. Nach einer gründlichen Übersicht über Piezoelektrizität und Schwingungsstrukturen werden die gebräuchlichsten self sensing Verfahren diskutiert, und das neue Verfahren wird mit den linearen Methoden theoretisch verglichen. Abschließend, nach einer kurzen Beschreibung der Elektronik, die für die reale Implementierung dieses Verfahrens notwendig ist, wird das neue self sensing Verfahren experimentell getestet und am Beispiel eines Kragträgers und einer Plattenstruktur mit einer häufig verwendeten linearen Methode verglichen. Die Experimente beweisen, dass die neue Methode im Unterschied zu linearen Verfahren eine gute Performance sowohl bei geringen als auch bei hohen Betriebsspannungen aufweist.

# Index

# 1.   Introduction

Vibrations are present everywhere in nature and they affect human life daily in many ways. Vibrations can be described as the oscillations of a system about an equilibrium point, and they can be periodic as well as random. Moreover, they can be desirable as well as undesirable. For example, the vibration generated by a guitar string produces sound, as well as in the case of loudspeakers, where the vibration of a membrane allows the reproduction of music. Even more, devices capable of generating vibrations are developed and integrated in many devices, as for example in mobile phones or gaming tools. In all these cases, vibrations are desirable mechanical phenomena. Nevertheless, in many other cases vibrations are not desirable, since they would waste energy, produce unwanted noise or damage and destroy systems affected by them. For example, the vibrations provided by a running engine are unwanted as they waste energy, or, in industrial environments, the vibrations generated by the machines result into acoustic pollution with a resulting discomfort for the workers. Finally, earthquakes are the typical case of undesired vibrations as it is well known about the damages it can provoke to civil buildings and thus to human beings. Thus, vibrations can be counter-productive and lead to discomfort or to danger.

Over the last years vibration control has become a fundamental research topic within the mechatronic area, in order to cope with undesired vibrations and to provide solutions to many practical applications. In particular, vibration control devices are classified as:

- passive control devices: the mechanical vibrating system is mechanically tuned to damp the vibrations within the frequency ranges of interest, and this is usually done by adding structural elements with the disadvantage of increasing the dimension, the weight and the complexity of the original system

- active control devices: they interact with the mechanical vibrating system by means of sensors and actuators capable of generating counter forces which are driven by a vibration controller so that no relevant additional mass is located on the system and the vibration damping capability depends strongly on the quality of the controller; such devices, anyways, require external power which must be provided for functioning

- hybrid control devices: such systems combine the features of passive and active control devices as they are able to convert the mechanical vibrations into electrical energy which is transferred to an electrical load which can dissipate it or store it; such devices are usually a good compromise in many applications as their damping capability is typically lower than active control devices.

Piezoelectric materials have found great application within the field of vibration control. The discovery of such materials is due to Jacques and Pierre Curies [CC80] that in the 1880 published a paper in the french scientific journal *Comptes Rendus* showing that some materials like Rochelle salt, zinc blende and others exhibit electrification when a pressure is applied. This effect is today also known as direct piezoelectric effect. The converse piezoelectric effect, which instead is the change in shape of these materials when an electric field is applied, has then been first theoretically predicted by Lippman and finally verified by the Curies brothers a year after [Mas81]. Because of these two effects, piezoelectric materials found great application in transducers, where their sensing (direct effect) and actuating (converse effect) capabilities are fundamental, although research work on piezoelectric materials has officialy started only after the end of the second world war. Nowadays piezoelectric materials are used in many applications, as detectors and generators of sonar waves, fuel injection systems, electronic drum pads transducers, loudspeakers, piezoelectric motors, inkjet printers and many more.

Moreover, piezoelectric materials found wide application in active and hybrid vibration control devices. Such materials, in fact, can actuate forces by electrical excitation on a very wide frequency range, making them suitable as actuators. Furthermore, they can sense deformations since they exhibit an electric voltage when deformed. Therefore such materials can be used as actuators as well as sensors in vibration control applications for modal wave and noise control.

Piezoelectric transducers are realized in different configurations, according to the way in which the deformation is required, which can be along a single axis or along a plane, as the ones which will be considered in this work. Such devices are simply made of a plate layer of piezoelectric ceramic covered by electrodes on both sides, and they are typically bonded on the vibrating structure to be controlled.

The first works presented in literature focused on the interaction between a piezoelectric plate actuator/sensor (transducer) and an elastic structure, as for example the works presented in 1986 by Anderson [And86] and in 1987 by Crawley and De Luis [CL87]. In both of these works, the authors study the static mechanical behaviour of a piezoelectric structure, i.e. a structure with integrated piezoelectric transducers, both in the case the transducers are surface mounted, i.e. bonded on the surface of the structure, or embedded, i.e. the piezoelectric plates are embedded within the structure. Moreover, piezoelectric transducers are also referred to as *distributed actuators and sensors*, as they actuate and sense over a defined spatial region of the structure, differently from point sensors and actuators which sense and actuate only over one point of the structure. A similar work was presented also by Lee in 1990 [Lee90], and extended and published in the same year in [LM90], where the author describes piezoelectric modal sensors and actuators, capable of sensing and actuating a single vibration mode of the structure. Indeed, each resonance frequency of an elastic structure is associated to a vibration mode, or spatial configuration of the structure at a certain frequency. Although it is a very interesting feature, it requires a well precise shaping of the piezoelectric electrodes according to the structural mode to be selected, which is not a practical procedure in some applications.

The research results on the static interaction between a piezoelectric transducer and an elastic structure have been the basic ground for the dynamic analysis of a piezoelectric structure for vibration control purposes. At this aim, Hagood presented in 1990 a fundamental work on the modelling of piezoelectric structure dynamics [HCF90], where, by means of the Rayleigh-Ritz formulation, a state space model is developed for the dynamic coupling between a structure and an electrical network through the piezoelectric effect. This modelling framework allows for several types of investigations, involving the fields of active vibration control and shunt damping (which is a hybrid vibration control). It is here important to remark that an elastic structure is usually modelled as a distributed parameter system, while in the derivation from Hagood it is possible to obtain a lumped parameter system model, as it will be shown in Chapter 2.

From the early 1990's until the present day, a variety of works on vibration control of elastic structures by means of piezoelectric actuators/sensors has been proposed and published ([PHS92], [MA10], [Urg10]). Since elastic structures are of different types, the most studied structures, also called standard structures, are wires, beams, rods, plates, shells and trusses as they are good approximations for many more complex structures.

Another important issue related to vibration control is about the optimal placement of sensors and actuators on a structure. In fact, the sensing and actuating capability of a piezoelectric transducer on a structure is strongly influenced by its position. Among the several approaches and solutions presented in literature, a remarkable contribution has been provided by Reza Moheimani in 2003 [RH03], in which an optimization approach for optimal placement of collocated piezoelectric actuators and sensors by means of modal and spatial controllability measures is proposed. Although optimal placement represents a very important aspect in vibration control, it will not be faced within this thesis work.

Vibration control loops make use of piezoelectric transducers both for actuating and for sensing. A very common choice is to use collocated actuator/sensor pairs, i.e. each piezoelectric actuator is coupled to a piezoelectric sensor which senses the same deformation as the piezoelectric actuator. Considering a planar structure, the two transducers would be located over the structure in the same position and one per side. This configuration allows an easier tuning of the vibration controllers, since the transfer function sensor vs. actuator can be modelled as a lumped parameter system. Thus, stability analysis becomes easier than in the case of non collocated actuator/sensor pairs. In fact, since structures are modelled as distributed parameter systems, the design of a stable vibration controller can become particularly burdensome.

In some applications it can be difficult or even impossible to realize a collocated actuator/sensor pair (space constraints limitations). In such cases, where transducer collocation is an issue, it can be advantageous to integrate the actuating and sensing capabilities in the same device, i.e., to use a piezoelectric material simultaneously as an actuator and as a sensor, thus realizing intrinsically a transducer collocation.

The concept of combining sensing and actuating features in the same device leads to the so-called self sensing techniques. A self sensing based system needs half of the amount of normally necessary transducers, making this approach appreciable for applications where miniaturization is desirable, since halving the number of transducers implies reducing the needed wiring, which is also a very important aspect for industrial applications where wiring can be burdensome and expensive.

Self sensing is applicable not only with piezoelectric materials, but in general with those materials that intrinsically hold information about the mechanical quantities such as force and displacement as well as about the electrical quantities, that in the case of piezoelectric materials are electric displacement and electric field strenght. Hence in a self sensing based control loop the mechanical quantities are reconstructed through the measurement of the electrical quantities and the model of the transducer. Piezoelectric materials perfectly match these requirements.

## 1.1. State of the art

One of the first works about self sensing piezoelectric technique was published by Dosch, Inman and Garcia in 1992 [DIG92]. In this paper the authors show that a piezoelectric transducer can simultaneously be used as an actuator and as a sensor by means of an electrical bridge circuit capable of measuring the strain of the transducer based on a reference capacitor whose capacitance is equal to the piezoelectric one. This technique was then experimentally verified by actively damping the vibration in a cantilever beam. Nevertheless, this kind of self sensing technique has its weak aspect in the reference capacitance chosen for balancing the electrical bridge.

In 1994 Anderson and Hagood [AH94] proposed the same technique and again verified it on a cantilevered beam as well as on a truss. Nevertheless, it is clearly remarked in this work how much difficulty is in the choice of the reference capacitance in order to match the structural dynamics. The authors wrote "*in contrast to the relatively easy matching of the system poles, matching of the modelled and measured zeros was extremely difficult to achieve. The chief difficulty was the hysteretic nature of the piezoelectric. This translated into a lossy capacitance that could not be matched by an ideally lossless reference capacitor*". This consideration will be more deeply investigated in Chapter 3 of this work, and represents the basic issue of all self sensing techniques.

In the same year Cole and Clark [CC94] proposed a solution for identifying the value of the reference capacitor needed for balancing the electrical bridge. By modelling the piezoelectric structure dynamics, the authors demonstrated that it is possible to identify the piezoelectric capacitance by driving it either with white noise or with a harmonic signal tuned at a structural resonance by using a LMS or a RLS algorithm. This result represents one of the milestones in the field of piezoelectric self sensing techniques as it provides an identification criterion for tuning the electrical bridges. Since this paper was exclusively theoretical, Vipperman and Clark published the first experimental results in 1995 and 1996 ([VC95] and [VC96a]). Not only the experimental results met the theoretical expectations, but they used the identification criterion for realizing adaptive identification of the piezoelectric capacitance, and proposed a hybrid analog and digital implementation of this technique.

The adaptive identification of the piezoelectric capacitance proposed by Vipperman and Clark can account for temperature variations and aging. Nevertheless, the difficulty of matching the piezoelectric capacitance was still not solved because of the hysteretic nature of piezoelectric materials. In fact, by increasing the driving voltage applied to the piezoelectric self sensing actuator, the mismatching between the piezoelectric capacitance and the reference capacitance leads to a considerable mismatch of the open loop response. Vipperman and Clark then proposed a modified technique in 1996 ([VC96b]) aiming at reducing the effects of the piezoelectric hysteretic nonlinearities by means of an artificial neural network. Unfortunately such technique needs to run offline and a lot of measurements are necessary, making this self sensing technique not suitable for practical applications.

Some papers tried to address the problem of matching the hysteretic nature of the piezoelectric capacitance, as for example Jones and Garcia in 1997 who proposed to use the piezoelectric constitutive equations together with hysteretic modelling [JG97]. Nevertheless they had given no indications about the identification criterion to be used for identifying the piezoelectric capacitance once that the transducer is bonded on the structure.

Small modifications to the basic linear technique were also proposed by Simmers et alter in 2004, who suggested to use capacitors in series and in parallel with the piezoelectric transducer to improve the balancing of the electrical bridge [SHM04].

Kuiper and Shitter [KS10] implemented the linear self sensing technique to damp the vibrations of an AFM scanner in order to increase its scanning speed and the quality of the obtained images. Moreover, the recent work [JQW11] brought the attention back to the problem of self sensing. In this work a neural network is used in the self sensing algorithm and the performance is checked by using an adaptive control algorithm.

## 1.2. Aim of this work

At the actual state of the art, all the proposed self sensing techniques are limited to the linear or quasi-linear working range of piezoelectric transducers since they fail at high driving voltages. In particular it is well known that piezoelectric materials exhibit nonlinear behaviours, which need to be modelled in order to have a reliable reconstruction of the mechanical quantities of interest. The most crucial nonlinearity is represented by the hysteresis, which is not considered in the mentioned references. Indeed, modelling the sensor/actuator electrical behaviour becomes more complex as the input voltage increases because of the nonlinearities of its characteristics, and this could lead to a degradation of the reconstruction quality which may even cause problems of instability when the reconstructed quantities are used as feedback for a vibration controller.

In this work a new self sensing reconstruction method is proposed that reaches good control performance even when consistent disturbances are acting on the structure. This circumstance requires the controller to drive the self sensing piezoelectric transducers at high voltages, where this material behaves in a strongly nonlinear way. The proposed self sensing technique, in fact, is based on a hysteretic model of the piezoelectric clamped capacitance which accounts for the piezoelectric nonlinearities, and sets no limits to the amplitude of the driving voltage. In order to prove the quality of this new self sensing technique, an experimental comparison with linear self sensing techniques is shown in order to highlight the main advantages that the approach proposed in this work is capable to offer.

Chapter 2 provides the necessary mathematical framework to analyse and discuss self sensing techniques. It presents the basic knowledge about piezoelectric transducers and their properties, as well as vibrating structures and their mathematical description. Finally, these two parts are joint together to describe the interaction of piezoelectric transducers with vibrating structures, and a mathematical model is provided. The information provided in this chapter is well known in literature, nevertheless this thesis work aims at presenting such scientific results in a common notation to provide a thorough scientific background to deal with piezoelectric self sensing techniques as presented in the following chapters.

Chapter 3 discusses self sensing techniques from the easiest formulations up to the adaptive algorithm proposed by Cole and Clark. Moreover, a new self sensing technique for large driving voltages is presented and theoretically explained. This technique, in fact, allows to use a self sensing actuator also at high driving voltages, and it is referred here as hysteretic reconstruction. Furthermore, different ways of implementing the afore mentioned self sensing techniques are presented, which can be analog, digital or hybrid implementations.

Chapter 4 discusses the electronics which is necessary for implementing self sensing techniques. In particular, the electronic circuits taken into consideration are a capacitance multiplier, a Sawyer Tower, an Active Sawyer Tower and an Active Sawyer Tower with temperature measurement. While the first circuitry allows to simulate a variable capacitance, the last three allow measuring the charge stored on a piezoelectric transducer while driving it.

Chapter 5 shows and discusses experimental results obtained to prove that the hysteretic reconstruction performs better than the linear one. In particular, positive position feedback and resonant controllers are tuned to damp the undesired vibrations of two standard structures, respectively a beam and a plate, by using a self sensing actuator. Linear and hysteretic reconstructions are implemented and the vibration control performance are compared and discussed.

Appendix A presents the Modified Prandtl-Ishlinskii model which has been used for modelling the hysteretic piezoelectric clamped capacitance, from its definition up to the identification algorithm.

Appendix B gives a short insight into the vibration controllers used in this work, which are a Positive Position Feedback (PPF) controller and a resonant controller.

# 2. Piezoelectricity and vibrating structures

This chapter is intended to provide a mathematical framework for piezoelectric vibrating structures (PVSs), i.e. elastic structures under vibrational excitation which are actuated and sensed by piezoelectric materials.

Section 2.1 deals with piezoelectricity. After presenting the physical effect, the linear constitutive equations and the typical nonlinear effects of a piezoelectric material are presented. Finally a short review of the most common piezoelectric transducers is shown.

Section 2.2 provides a mathematical framework to study and analyze vibrating elastic structures. In particular, after introducing the basics of modal analysis, two elastic standard structures are taken into consideration: a cantilevered beam and a simply-supported plate.

Section 2.3 takes into consideration the interaction of a piezoelectric transducer with a vibrating structure, both in the case of static and dynamic interaction. In particular, a state-space model for PVS is shown, and applied again to a cantilevered beam and to a simply-supported plate. Moreover, the relative transfer functions for a PVS are retrieved.

## 2.1. Piezoelectricity

*Piezoelectric* materials are a subset of *dielectric* materials. In particular such materials exhibit a charge when a deformation is applied. Within the set of piezoelectric materials, the ones who have a spontaneous polarization are called *pyroelectric* materials, which exhibit a charge also when a change in temperature occurs. Finally, a subset of pyroelectric materials is represented by *ferroelectric* materials, whose spontaneous polarization can be easily changed by applying an external electrical field. Such materials exhibit a domain structure and a spontaneous polarization at temperatures below the Curie point.

### 2.1.1. Piezoelectric effect

The piezoelectric effect was investigated in 1880 by Jacques and Pierre Curie [CC80] while Pierre was investigating the relationship between pyroelectricity (exhibition of surface charge corresponding to temperature changes) and crystal symmetry. By studying different materials like quartz, cane sugar and Rochelle salt, they were able to demonstrate that such materials exhibit charge as response to an applied stress. This effect is called *direct piezoelectric effect* (Figure 2.1a). Modern materials can generate up to some thousands volts due to this effect. The Curie brothers demonstrated also that a change in dimensions of the crystal can be obtained by applying an electric field, that is the *converse (or inverse) piezoelectric effect* (Figure 2.1b). Such work was also predicted theoretically by Lippman based on thermodynamic principles and then verified by the Curies in the following years.

Both piezoelectric effects are a consequence of the nonsymmetric nature of piezoelectric materials (ceramics and polymers), which also enables the dipole switching which leads to hysteresis and constitutive nonlinearities inherent to ferroelectric and piezoelectric materials at all drive levels. Direct and converse piezoelectric effects can be then used respectively for sensing and actuation purposes.

After the first investigations on the piezoelectric effect in the late 19th century, research focused mostly on piezoceramic compound like the $BaTiO_3$ (barium titanate), belonging to the perovskite family of piezoceramic materials and served as a precursor to lead zirconate titanate (PZT) which is at the moment one of the most common and used piezoceramic.

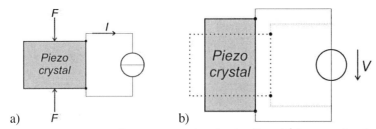

**Figure 2.1 – Piezoelectric effect. (a) Direct piezoelectric effect. (b) Converse piezoelectric effect.**

These materials are characterized by a high dielectric and piezoelectric strength and can operate in a broad range of temperature. They are also characterized by a high rigidity, which is a very favourable feature for force actuators and a drawback for deformation sensors. Polyvinilidene fluoride (PVDF) is a soft piezoelectric material which is characterized by a low rigidity and a high piezoelectric coefficient and consequently it is often a common choice for sensor applications.

### 2.1.2. Constitutive equations for piezoelectric materials

An initial linear characterization of the electromechanical properties of piezoelectric materials was published in [Voi10]. Even though piezoelectric materials are typically nonlinear, for low to moderate input fields they provide approximately linear responses, which motivate the formulation of linear constitutive equations. Let us then consider the thermodynamic formulation for a piezoelectric material, which consider mechanical, electrical and thermal processes as shown in Figure 2.2.

As one can see, many physical effects interact at the same time for the behaviour of a piezoelectric material. In particular, the intensive variables electric field strength $E$, stress $\sigma$ and temperature $T$ are disposed at the vertices of the outer triangle, while the extensive variables electric displacement field $D$, strain $S$ and entropy $H$ are represented in the inner triangle.

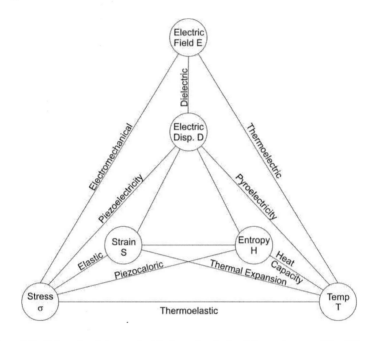

*Figure 2.2 – Interaction between electrical, mechanical and thermal process [New05].*

A linear set of constitutive equations for a piezoelectric material can then be derived and expressed as:

$$\Delta H = \left( C^{\sigma,E} / T_R \right) \Delta T + \alpha_{ij}^E \sigma_{ij} + p_i^\sigma E_i$$
$$S_{ij} = \alpha_{ij}^E \Delta T + s_{ijkl}^{E,T} \sigma_{kl} + d_{nij}^T E_n \qquad (2.1)$$
$$D_n = p_n^\sigma \Delta T + d_{nij}^T \sigma_{ij} + \varepsilon_{nm}^{\sigma,T} E_m$$

where each physical effect is approximated with first-order static relationships, and $\Delta T = T - T_R$, $\Delta H = H - H_R$ are the change of temperature and entropy, and $T_R$ and $H_R$ are the reference temperature and entropy, respectively. In particular, in Table 2.1 a list of the effects and their relative coefficients is shown.

| Physical effect | Symbol |
|---|---|
| Specific heat | $C^{\sigma,E}/T$ |
| Thermal expansion and piezocaloric effect | $\alpha_{ij}^{E}$ |
| Pyroelectricity and electrocaloric effect | $p_{i}^{\sigma}$ |
| Elastic compliance | $s_{ijkl}^{E,T}$ |
| Direct and converse piezoelectric effect | $d_{nij}^{T}$ |
| Permittivity | $\varepsilon_{nm}^{\sigma,T}$ |

*Table 2.1 – Physical effects and coefficients.*

The apexes of each coefficient indicate the physical quantities which are constant while measuring the coefficient itself. For example, $\varepsilon_{nm}^{\sigma,T}$ is the permittivity measured at constant stress $\sigma$ and constant temperature $T$, while the subscripts $n,m$ are the cartesian coordinates. Consequently, $s^{E,T}$ is a fourth rank compliance tensor measured at constant electric field and temperature.

The number of coefficients describing the overall behaviour is relatively high. Nevertheless, it can be significantly reduced considering the elastic symmetries as well as electric symmetries due to poling. In fact, the stress and strain tensors are symmetric, leading to $\sigma_{ij} = \sigma_{ji}$ and $s_{ij} = s_{ji}$. This implies, for example, that $d_{nij}$ is symmetric in $i$ and $j$. In order to formulate the system as a matrix equation, the indices $ij$ are replaced with a single index $m$ as shown in Table 2.2, according to the Voigt notation.

| Tensor $(i,j)$ | 11 | 22 | 33 | 23,32 | 31,13 | 12,21 |
|---|---|---|---|---|---|---|
| Matrix $(m)$ | 1 | 2 | 3 | 4 | 5 | 6 |

*Table 2.2 – Voigt notation for symmetric tensors reduction.*

Finally the coefficients $d_{nij}$ can be written as $d_{nij} = d_{nm}$ for $m = 1, 2, 3$ and $d_{nij} = \frac{1}{2} d_{nm}$ for $m = 4, 5, 6$. In Figure 2.3 a graphical explanation of the Voigt notation is shown. Similar conventions are applied for the others coefficients. These convention results are very useful to have a compact description of the coefficients. For example, for the coefficient $d_{nm}$ the index $n$ indicates the direction of the field $E$ while the index $m$ the direction of the strain $S$ [Smi05]. A further reduction of the order can be obtained by considering that a piezoelectic material is often poled in the direction 3. Moreover, a PZT material is also isotropic. In such case all the piezoelectric coefficients are zero except for $d_{33}$, $d_{31} = d_{32}$ and $d_{15} = d_{24}$.

The linear constitutive equations in (2.1) represent a set of 3 equations in 6 variables. More common formulations instead use a set of 2 equations in 4 variables, in particular only the mechanical and the electrical ones. In fact, in many applications temperature related aspects can be neglected. By neglecting in fact the thermal variables, equations (2.1) reduce to the matrix equation

$$
\begin{aligned}
\mathbf{S} &= \mathbf{s}^E \boldsymbol{\sigma} + \mathbf{d}_t \mathbf{E} \\
\mathbf{D} &= \mathbf{d}\boldsymbol{\sigma} + \boldsymbol{\varepsilon}^\sigma \mathbf{E},
\end{aligned}
\tag{2.2}
$$

where the pedix $t$ indicates the transposition of the matrix and the stress $\boldsymbol{\sigma}$ and electric field $\mathbf{E}$ are the independent variables. This equation can also be written extensively as

$$
\begin{bmatrix}
S_1 \\ S_2 \\ S_3 \\ S_4 \\ S_5 \\ S_6 \\ \hline D_1 \\ D_2 \\ D_3
\end{bmatrix}
=
\begin{bmatrix}
s_{11}^E & s_{12}^E & s_{13}^E & 0 & 0 & 0 & 0 & 0 & d_{31} \\
s_{12}^E & s_{11}^E & s_{13}^E & 0 & 0 & 0 & 0 & 0 & d_{31} \\
s_{13}^E & s_{13}^E & s_{33}^E & 0 & 0 & 0 & 0 & 0 & d_{33} \\
0 & 0 & 0 & s_{44}^E & 0 & 0 & 0 & d_{15} & 0 \\
0 & 0 & 0 & 0 & s_{44}^E & 0 & d_{15} & 0 & 0 \\
0 & 0 & 0 & 0 & 0 & s_{66}^E & 0 & 0 & 0 \\
\hline
0 & 0 & 0 & 0 & d_{15} & 0 & \varepsilon_{11}^\sigma & 0 & 0 \\
0 & 0 & 0 & d_{15} & 0 & 0 & 0 & \varepsilon_{22}^\sigma & 0 \\
d_{31} & d_{31} & d_{33} & 0 & 0 & 0 & 0 & 0 & \varepsilon_{33}^\sigma
\end{bmatrix}
\begin{bmatrix}
\sigma_1 \\ \sigma_2 \\ \sigma_3 \\ \sigma_4 \\ \sigma_5 \\ \sigma_6 \\ \hline E_1 \\ E_2 \\ E_3
\end{bmatrix}.
\tag{2.3}
$$

According to the set of independent variables which are chosen, four different sets of linear constitutive equations can be written. In particular, at the aim of this work, it is crucial to report at least another formulation, which considers the strain $S$ and the electric field $E$ as independent variables:

$$\sigma = c^E S - e_t E$$
$$D = eS + \varepsilon^S E,$$
(2.4)

where $c^E$ is the matrix of Young's modulus measured at constant electric field strength and $e$ is the piezoelectric constant matrix.

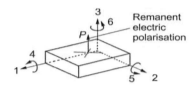

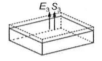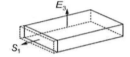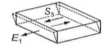

*Figure 2.3 – Definition of the axes in piezo materials. The digits 4, 5 and 6 indicate the shear on the axes 1, 2 and 3.*

In both formulations shown in (2.2) and (2.4), the first equation is called *actuator equation*, while the second one *sensor equation*. In fact, the first one is used to describe the actuated displacement or actuated stress of a piezoelectric material, while the second one the generated electric displacement resulting from an external force or deformation.

As one can see, the linear constitutive equations shown in (2.2) make use of the dielectric permittivity matrix measured at constant stress $\varepsilon^\sigma$, also called *unconstrained permittivity*, while in equation (2.4) the dielectric permittivity measured at constant strain $\varepsilon^S$, called *clamped permittivity*, is used.

Since typically piezoelectric datasheets do not provide the $e$ constant matrix and the clamped dielectric matrix $\varepsilon^S$, they can be derived from the other parameters by using the following formula:

$$\mathbf{e} = \mathbf{dc}^E \tag{2.5}$$

$$\mathbf{\varepsilon}^S = \mathbf{\varepsilon}^\sigma - \mathbf{dc}^E \mathbf{d}_t , \tag{2.6}$$

which are easy to obtain by algebraic manipulation.

Finally it can be of interest to remark that other linear constitutive equations used in other works can refer to the polarization density $P$ rather than to the electric displacement field $D$. Nevertheless it is possible to change from one notation to the other one by using the following formula:

$$\mathbf{D} = \varepsilon_0 \mathbf{E} + \mathbf{P} . \tag{2.7}$$

## 2.1.3. Piezoelectric nonlinearities

The linear constitutive equations shown in (2.1), (2.2) and (2.4) allow to create a framework useful to describe in a static first-order approximation the behaviour of a piezoelectric material. Nevertheless piezoelectric materials exhibit non-static nonlinearities.

### *Hysteretic behaviour*

In particular the dipole switching mechanisms produce hysteresis in the relation between the electric input field $E$ and stress $\sigma$ and strain $S$.

Typical hysteresis loops for a piezoelectric material are shown in Figure 2.4 where $P_r$ and $S_r$ stand for remanent polarization and permanent strain, respectively, and $E_c$ is the coercive field strength. Such field can be defined as the electric field required to oppose and bring to zero the internal polarization. Figure 2.4a relates the polarization density $P$ to the electric field $E$, eventhough a similar diagram can be obtained by considering the electric displacement $D$, since it differs only by the term $\varepsilon_0 E$. Moreover the hysteresis shown in Figure 2.4b is also called *butterfly trajectory* due to its shape and it is typical for piezoelectric materials. In the first diagram one can observe the saturation effect which limits the behaviour, especially for actuating purposes, where the strain is saturated at high electric field strengths.

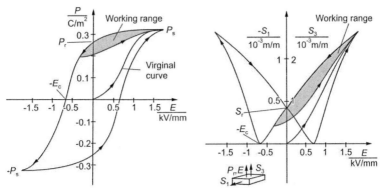

**Figure 2.4 – Hysteretic behaviour. (a) Electric field-polarization hysteresis. (b) Electric field-strain hysteresis [Jan04].**

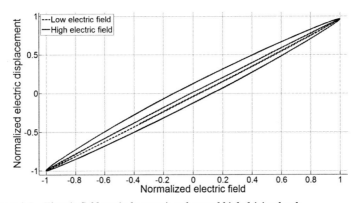

**Figure 2.5 – Electric field-strain hysteresis at low and high driving levels.**

Typical piezoelectric applications do not excite the piezoelectric material in its full driving range, but in a smaller range, also called *working range*. In such range the hysteretic curve becomes less complex, and it becomes even more linear by decreasing the electric field range of excitation, in a way that the linear constitutive equations can describe quite precisely the real behaviour of the material (see Figure 2.5).

## Creep effect

Another important nonlinear effect of piezoelectric material is *creep*, which in material science is the tendency of a solid material to move slowly or to deform permanently under the influence of stress. Piezoelectric materials exhibit a creep effect of the electric displacement field and of the strain due to a step change of the electric field strength (see Figure 2.6). In particular creep always occurs in the same direction as the dimensional change produced by the electric field step. Typical values range are from 1% to 20% of the expected elongation with a time constant between 10 and 100 seconds.

Creep effect has a decreasing temporal behaviour described by

$$S(t) = S(T_s) + \gamma \log\left(\frac{t}{T_s}\right), \tag{2.8}$$

where $S$ is the strain, $\gamma$ a weight factor of the creep effect and $T_s$ the sampling time [Kuh01].

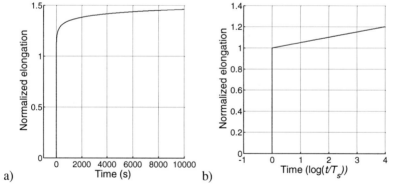

a)                                                                 b)

**Figure 2.6 – Creep of the normalized elongation after a change in rate of the electric field strength, with $T_s=1$, $\gamma=0.5$ and $S(T_s)=1$. (a) Linear time scale. (b) Logarithmic time scale.**

Thus, due to the presence of creep, piezoelectric materials exhibit rate dependent hysteretic characteristics. In particular, it is more relevant at the aim of this work to consider the creep effect for its frequency dependent behaviour.

As one can see in Figure 2.7, the amplitude peak-to-peak of a hysteretic characteristic with creep decreases exponentially over the frequency. Such property is of interest for certain dynamic applications, since the characteristic changes quickly at low frequencies to become more stable at higher frequencies.

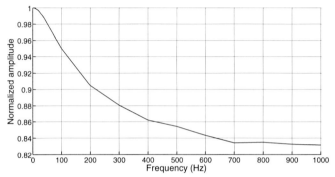

*Figure 2.7 – Normalized amplitude of a piezoelectric hysteretic characteristic with creep.*

### *Power dissipation and temperature issues*

Piezoelectric materials behave electrically like capacitors. In particular, at temperatures well below the Curie temperature, the internal (or leakage) resistance is particularly high, approximately in the range of 1 GΩ … 1 TΩ. Consequently in static operation the current flowing through the piezoelectric material is typically in the range of 1 pA … 1nA and thus it can be mostly neglected. That means that nearly no power is required or needed to maintain a certain state of activation. The needed power should theoretically be only related to the processes of charging and discharging. Typical capacitors do not dissipate energy via these processes, but this is not valid for piezoelectric materials. In fact, they dissipate energy at each charging/discharging cycle due to its nonlinear nature. In fact, the typical hysteretic behaviour of piezoelectric materials is the result of the internal friction among the dipoles which move according to the applied electric field. By considering a piezoelectric material with a capacitance $C$, driven at a voltage $V$ and at a frequency $f$, the dissipated power $P$ can be calculated as

$$P = 2\pi f C \tan(\delta) V^2 , \tag{2.9}$$

where $\tan \delta$ is the dielectric loss factor, which is usually measured at low voltages at a frequency of 1 kHz. This factor is usually higher for soft piezoelectric materials, in the range of 2% to 4%, while hard piezoelectric materials have a dielectric loss factor of 0.5%.

The dissipated power determines an increase in temperature, which is dependent on the heat capacity of the piezoelectric material and on which means are adopted to transfer the heat to the surrounding. Nevertheless, a variation of the temperature affects the characteristics of the piezoelectric material. As one can see from equation (2.1), three coefficients are temperature dependent, and they are:

- the elastic compliance $\mathbf{s}^{E,T}$,
- the piezoelectric charge constant $\mathbf{d}^T$,
- permittivity $\mathbf{\varepsilon}^{\sigma,T}$.

In particular, let us consider the *coupling coefficient*

$$k_{33} = \frac{d_{33}^T}{\sqrt{s_{33}^{E,T} \varepsilon_{33}^{\sigma,T}}} , \tag{2.10}$$

which is defined as the square root of the ratio between the electrical energy stored in the 3$^{rd}$ direction and the mechanical energy applied in the 3$^{rd}$ direction. As shown in [MTA07], such coefficient decreases down to zero by varying the piezoelectric material temperature from the room temperature up to the Curie temperature. In fact, piezoelectric materials (as well as ferroelectric materials) become paraelectric at temperatures above the Curie point. This limiting temperature is then specific for each material compound.

At the aim of this work, it is also of interest to analyze the permittivity temperature behaviour. In fact, it is important to consider that such value increases over the temperature. PZT ceramics can variate their capacitance of about 60 ... 80% on a range between the room and the Curie temperatures. In self sensing applications this is a critical aspect, since strong changes in the piezoelectric capacitance due to temperature changes must be taken into consideration. Anyways detailed information per each kind of piezoelectric material are usually provided by the producer.

## 2.1.4. Piezoelectric transducers and actuators

Piezoelectric transducers can be classified basing on the piezoelectric effect which is used. In particular, we can list them based on the $d_{33}$, $d_{31}$ and $d_{15}$ effects.

### *Stack translators*

Piezoelectric stack actuators are the most common piezoelectric transducers, which are composed of many stacked piezoelectric discs, provided with electrodes, glued together. The discs are connected each other in opposing polarization. Such actuators use the $d_{33}$ effect, consequently providing a high elongation in the polarization direction. Each disc has a thickness between 0.3 and 1 mm [Jan07], which allows to drive these actuators up to a voltage between 600 V and 2000 V, since the typical maximum electric field for a piezoelectric ceramic is 2000 V/mm.

Multilayer actuators have a similar design, but have the advantage of providing the required electric field strength at lower voltages. In fact, such actuators are made of piezoelectric slices which thickness ranges from 60 μm up to 80 μm. Consequently, they can be driven up to a maximum voltage which is in the range between 120 V and 160 V.

The $d_{33}$ effect based actuators are typically used for high force generation or for precise positioning applications.

### Laminar translators

These kind of actuators make use of the $d_{31}$ effect, that means the desired elongation is exhibited on the perpendicular plane of the polarization direction, i.e. the $1^{st}$ and $2^{nd}$ axes of the material. PZT ceramic are isotropic, consequently the material strains equally in both directions, while PVDF polymers have a typical ratio of 5:1 between the $1^{st}$ and the $2^{nd}$ direction, since they are anisotropic materials. Nevertheless, PVDF materials are usually used for sensing due to their low actuation capability.

A typical ceramic provided with electrodes has a thickness between 0.1 mm up to 1 mm, with a driving voltage ranging between 200 V and 2000 V. Such actuators have found great application in vibration control, since they are usually connected to structures for vibration control purposes, since they can act as actuators as well as sensors.

### Shear elements

These actuators are based on the $d_{15}$ effect, which exhibits a shear motion of the ceramic about its $2^{nd}$ axes. Such transducers have been typically used as sensors, even though recent applications make use of this effect for particular positioning systems as the ones used in inkjet head printers.

### Piezoelectric actuators: examples

The piezoelectric transducers listed above have been the main core of unconventional actuators development. Among the wide variety of actuators, like inchworm motors, ultrasonic motors, displacement amplifiers and so on, in this section the so-called *Pendulum Actuator* (Figure 2.8) is briefly shown and discussed. Such actuator has been designed and developed at the LPA department, Saarland University, Germany, in order to produce high dynamic forces even though in a narrow frequency range (50 Hz up to 100 Hz range) [MJG09].

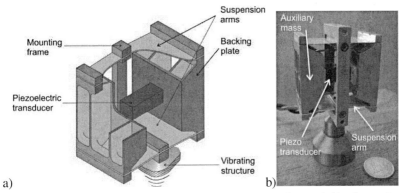

**Figure 2.8 – Pendulum Actuator. a) 3d model representation and b) photo of a demonstration device [GMJ11].**

Such device comes ideally from the cathegory of active mass dampers, where an auxiliary mass is connected to the structural frame via some elastic suspensions and an active material drives its motion. In the Pendulum Actuator the active material is a piezoelectric stack translator. The originality of this device is in its kinematic, which makes the device operating in the so-called *pendulating motion* that can be considered a nonlinear resonance of the device. This feature allows the generation of high forces on the structure to which it is connected, even though affected by the distortion introduced by the nonlinear kinematic. Nevertheless some research work has been done in order to decrease the amount of distortion introduced in the generated forces [GMJ12].

## 2.2. Vibrating structures

In this section standard structures such as beams and plates are introduced and presented by means of the so-called *modal analysis*, which provides a mathematical background for the study of the dynamic properties of structures under vibrational excitation. Sections 2.2.1-2.2.3 are taken from [RHF03], where a deeper analysis can be found, and here reported for completeness.

## 2.2.1. Modal analysis

Modal analysis is a mathematical background needed for analyzing and studying the dynamic properties of structures which are under vibrational excitation [RHF03]. Let us consider the following partial differential equation (PDE):

$$\mathcal{L}\{y(t,r)\} + \mathcal{M}\left\{\frac{\partial^2 y(t,r)}{\partial t^2}\right\} = f(t,r),\qquad(2.11)$$

where $r$ and $t$ are respectively the space and time variables and the space is defined over a domain $\mathcal{R}$, $\mathcal{L}$ and $\mathcal{M}$ are linear homogeneous differential operators respectively of order $2p$ and $2q$, with $q \leq p$, and $f$ and $y$ are respectively the system input and output spatially distributed over the domain $\mathcal{R}$. The equation (2.11) is defined based on the following boundary conditions:

$$B_l\{y(t,r)\} = 0,\qquad(2.12)$$

with $l = 1,2,\dots,p$. As one can see, equation (2.11) describes spatially and temporally the dynamic behaviour of $y$. Modal analysis is concerned with finding a solution to equation (2.11) of the form

$$y(t,r) = \sum_{i=1}^{\infty} \Phi_i(r) q_i(t),\qquad(2.13)$$

where $\Phi_i(r)$ are the eigenfunctions which are obtained by solving the eigenvalue problem associated to (2.11). That is,

$$\mathcal{L}\{\Phi_i(r)\} = \lambda_i \mathcal{M}\{\Phi_i(r)\}\qquad(2.14)$$

with the associated boundary conditions,

$$\mathcal{B}_l\{\Phi_i(r)\} = 0 \, ,$$

where $l=1,2,\dots,p$ and $i=1,2,\dots,\infty$.

The eigenvalue problem has a solution made of an infinite set of eigenvalues $\lambda_i$ associated to eigenfunctions $\Phi_i(r)$, which are also called *modal shapes* or *mode shapes*. Each eigenvalue is also related to a natural angular frequency $\omega_i$ of the system, according to this expression:

$$\lambda_i = \omega_i^2 \, . \qquad (2.15)$$

Assuming the $\mathcal{L}$ operator to be self-adjoint and positive definite, one can obtain the following properties:

$$\lambda_1 \le \lambda_2 \le \dots \qquad (2.16)$$

and

$$\int_{\mathcal{R}} \Phi_i(r) \mathcal{L}\{\Phi_j(r)\} dr = \delta_{ij} \omega_i^2 \, , \qquad (2.17)$$

$$\int_{\mathcal{R}} \Phi_i(r) \mathcal{M}\{\Phi_j(r)\} dr = \delta_{ij} \, , \qquad (2.18)$$

where $\delta_{ij}$ is the Kronecker delta function, that means $\delta_{ij} = 1$ for $i = j$, and equations (2.17) and (2.18) are called orthogonality conditions.

By substituting equation (2.13) in (2.11), multiplying both sides by $\Phi_j(r)$, integrating over the domain $\mathcal{R}$, and using the orthogonality conditions, one can get an infinite number of decoupled second order ordinary differential equations:

$$\ddot{q}_i(t) + \omega_i^2 q_i(t) = Q_i(t) \qquad (2.19)$$

where

$$Q_i(t) = \int_{\mathcal{R}} \Phi_i(r) f(t,r) \, dr \, . \tag{2.20}$$

Let us assume that $Q_i(t)$ can be written as

$$Q_i(t) = F_i u(t), \tag{2.21}$$

which means that $f(t,r)$ can be decomposed in its spatial and temporal components, and $u(t)$ is the input of the system. Starting from equation (2.19) one can obtain the following transfer function by using the Laplace transform:

$$G(s,r) = \sum_{i=1}^{\infty} \frac{\Phi_i(r) F_i}{s^2 + \omega_i^2} \, . \tag{2.22}$$

Standard elastic vibrating structures such as beams or plates are governed by partial differential equation (PDE) like the one in equation (2.11).

## 2.2.2. Modal analysis of a thin cantilevered beam

The Euler-Bernoulli formulation of a thin beam will be presented here by using the modal analysis approach discussed in the previous section. In particular, we make the following assumptions:

- the material respects the Hooke's law,
- the shear deformation is negligible with respect to the bending deformation,
- the rotation of the element is negligible compared to the vertical translation.

Let us then consider a thin (the thickness is some order of magnitude smaller than the length and the width) beam of length $L$, with mass density

$\rho(x)$, Young's modulus $E$ and second moment of area $I$, characterized by a flexural rigidity $EI(x)$, where a distributed force $f(x)$ is acting on it and where $x$ is the spatial coordinate and $w$ the vertical translation along the $y$ axes, as shown in Figure 2.9.

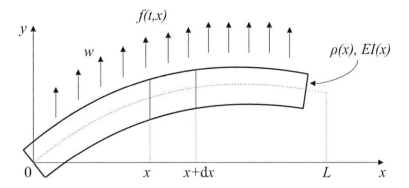

*Figure 2.9 – Flexural vibration of a beam.*

Let us then consider the free-body diagram of an element of the beam $dx$ as shown in Figure 2.10, where $Q$ and $M$ denote the shearing force and bending moment. Applying Newton's second law to the vertical force components, we obtain:

$$\left(Q(t,x) + \frac{\partial Q(t,x)}{\partial x}dx\right) - Q(t,x) + f(t,x)dx = \rho A(x)dx \frac{\partial^2 w(t,x)}{\partial t^2}, \quad (2.23)$$

which can be rewritten as

$$\frac{\partial Q(t,x)}{\partial x} + f(t,x) = \rho A(x)\frac{\partial^2 w(t,x)}{\partial t^2}, \quad (2.24)$$

where $A(x)$ is the section of the beam.

Moreover, let us consider the moment equilibrium equation about the $z$-axis (out-of-plane direction):

$$\left(M(t,x)+\frac{\partial M(t,x)}{\partial x}dx\right)-M(t,x)+$$
$$+\left(Q(t,x)+\frac{\partial Q(t,x)}{\partial x}dx\right)dx+f(t,x)dx\frac{dx}{2}=0.\qquad(2.25)$$

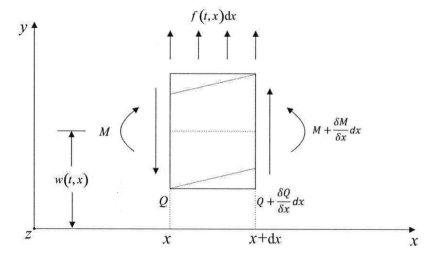

*Figure 2.10 – A small element of the beam.*

Then, by cancelling the higher order $dx$ terms and by simplifying, we can get the expression for the shearing force in terms of bending moments:

$$Q(t,x)=-\frac{\partial M(t,x)}{\partial x}.\qquad(2.26)$$

Furthermore, the bending moment can be related to the curvature of the element by using the following expression:

$$M(t,x)=EI(x)\frac{\partial^2 w(t,x)}{\partial x^2}.\qquad(2.27)$$

Finally, combining equation (2.24), (2.26) and (2.27), we can get the Euler-Bernoulli equation for a thin beam:

$$\frac{\partial^2}{\partial x^2}\left( EI\left(x\right)\frac{\partial^2 w(t,x)}{\partial x^2}\right) + \rho A(x)\frac{\partial^2 w(t,x)}{\partial t^2} = f(t,x)\,, \qquad (2.28)$$

which resembles equation (2.11).

After defining the Euler-Bernoulli equation for a thin beam, some ideal boundary conditions can be defined as follows:

- **Clamped end at** $x = x_0$**:** the deflection and slope of the curve are zero:

$$w(t,x_0) = 0, \qquad \frac{\partial w(t,x)}{\partial x}\bigg|_{x=x_0} = 0\,. \qquad (2.29)$$

- **Hinged (pinned) end at** $x = x_0$**:** The deflection and bending moment are zero.

$$w(t,x_0) = 0, \qquad EI\left(x\right)\frac{\partial^2 w(t,x)}{\partial x^2}\bigg|_{x=x_0} = 0\,. \qquad (2.30)$$

- **Free end at** $x = x_0$**:** The shearing force and bending moment are zero.

$$\frac{\partial}{\partial x}\left( EI\left(x\right)\frac{\partial^2 w(t,x)}{\partial x^2}\right)\bigg|_{x=x_0} = 0, \qquad EI\left(x\right)\frac{\partial^2 w(t,x)}{\partial x^2}\bigg|_{x=x_0} = 0\,. \qquad (2.31)$$

The above boundary conditions are defined for a generic point $x_0$ of the structure, which is usually associated to the points 0 or $L$, i.e. the two extremes of the beam.

Let us then consider a thin beam which is clamped on one side and free at the other one, with flexural rigidity, density and cross section constant over the spatial variable $x$, as shown in Figure 2.11. Moreover a point force $u(t)$ is acting on the beam at a point $x = x_1$.

The Euler-Bernoulli equation, shown in (2.28), is a fourth order differential equation, consequently it needs four boundary conditions in order to be solved, that in the case of a cantilevered beam are:

$$w(t,0) = 0$$
$$\left.\frac{\partial w(t,x)}{\partial x}\right|_{x=0} = 0$$
$$\left.\frac{\partial^2 w(t,x)}{\partial x^2}\right|_{x=L} = 0 \qquad\qquad (2.32)$$
$$\left.\frac{\partial^3 w(t,x)}{\partial x^3}\right|_{x=L} = 0.$$

According to the notation shown in section 2.2.1, we can write that:

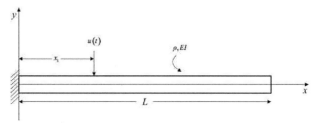

*Figure 2.11 – A cantilevered beam.*

$$\mathcal{L} = EI\frac{\partial^4}{\partial x^4},$$

$$\mathcal{M} = \rho A,$$

$$f(t,x) = u(t)\delta(x - x_1), \qquad\qquad (2.33)$$

where $\mathcal{R} = [0,L]$. Let us define two subsets of the domain $\mathcal{R}$, respectively

$S_0 = \{x \in \mathcal{R} \| x = 0\}$ and $S_L = \{x \in \mathcal{R} \| x = L\}$, one can define the following boundary conditions per each subset as:

$$\mathcal{B}_{10} = 1,$$

$$\mathcal{B}_{20} = \frac{\partial}{\partial x},$$

$$\mathcal{B}_{1L} = \frac{\partial^2}{\partial x^2},$$

$$B_{2L} = \frac{\partial^3}{\partial x^3}. \tag{2.34}$$

Finally, also the orthogonality conditions can be expressed as:

$$\int_0^L \Phi_i(x)\Phi_j(x)\rho A \mathrm{d}x = \delta_{ij}, \tag{2.35}$$

$$\int_0^L \Phi_i(x)\Phi_j''(x) EI \mathrm{d}x = \omega_i^2 \delta_{ij}, \tag{2.36}$$

where, recalling equation (2.14), $\omega_i$ are the solutions to the following eigenvalue problem:

$$\Phi_i''(x) - \lambda_i^4 \Phi_i(x) = 0, \tag{2.37}$$

where by definition

$$\lambda_i^4 = \frac{\rho A \omega_i^2}{EI}. \tag{2.38}$$

The eigenfunctions $\phi_i(x)$ have to satisfy the following boundary conditions:

$$\begin{cases} \Phi_i(0) = 0 \\ \Phi_i'(0) = 0 \\ \Phi_i''(L) = 0 \\ \Phi_i'''(L) = 0. \end{cases} \qquad (2.39)$$

A general solution for equation (2.37) is of the form:

$$\phi_i(x) = A_i \sin(\lambda_i x) + B_i \cosh(\lambda_i x) + C_i \sinh(\lambda_i x) + D_i \cos(\lambda_i x). \qquad (2.40)$$

The first boundary condition of equation (2.39) leads to $B_i + D_i = 0$, while the second one leads to $A_i + C_i = 0$. The solution of equation (2.40) can then be rewritten as:

$$\phi_i(x) = A_i \left( \sin(\lambda_i x) - \sinh(\lambda_i x) \right) + B_i \left( \cos(\lambda_i x) - \cosh(\lambda_i x) \right). \qquad (2.41)$$

The last two boundary conditions lead instead to the two following equations:

$$A_i \left( \sin(\lambda_i L) + \sinh(\lambda_i L) \right) + B_i \left( \cos(\lambda_i L) + \cosh(\lambda_i L) \right) = 0, \qquad (2.42)$$

$$A_i \left( \cos(\lambda_i L) + \cosh(\lambda_i L) \right) - B_i \left( \sin(\lambda_i L) - \sinh(\lambda_i L) \right) = 0. \qquad (2.43)$$

By considering that $\cosh(\lambda_i L)^2 - \sinh(\lambda_i L)^2 = 1$, that $\cos(\lambda_i L)^2 + \sin(\lambda_i L)^2 = 1$ and that $A_i \neq 0$ in order to get a nontrivial solution, we can obtain the following equation:

$$\cos(\lambda_i L)\cosh(\lambda_i L) = -1. \qquad (2.44)$$

This transcendental equation can be solved by using numerical methods and has an infinite number of solutions, whose some of them are shown in Table 2.3.

| Index $i$ | $\lambda_i L$ |
|---|---|
| 1 | 1.8751 |
| 2 | 4.6940 |
| 3 | 7.8547 |
| 4 | 10.9955 |

*Table 2.3 – First four solutions of equation (2.44).*

Finally we can get the normalized mode shapes, i.e. $A_i = 1$:

$$\phi_i(x) = \left(\sin(\lambda_i x) - \sinh(\lambda_i x)\right) - \alpha_i \left(\cos(\lambda_i x) - \cosh(\lambda_i x)\right), \quad (2.45)$$

where

$$\alpha_i = \frac{\sin(\lambda_i L) + \sinh(\lambda_i L)}{\cos(\lambda_i L) + \cosh(\lambda_i L)}. \quad (2.46)$$

Mode shapes play a key rule in the spatially distributed model of a beam. Let us consider a thin cantilevered beam whose parameters are shown in Table 1.4. The first four mode shapes are shown in Figure 2.12.

| Parameters | Symbol | Value |
|---|---|---|
| Length | $L$ | 600 mm |
| Width | $b$ | 50 mm |
| Thickness | $h$ | 3 mm |
| Young modulus | $E$ | 70 GPa |
| Density | $\rho$ | 2700 kg/m$^3$ |
| Poisson ratio | $\upsilon$ | 0.3 |

*Table 2.4 – Parameters of the cantilevered aluminum beam.*

Finally, by considering the Laplace transform of the applied force $u(t)$ and the transverse deflection of the beam $z(t,r)$, respectively $u(s)$ and $y(s,r)$, we can get the transfer function between these two quantities as:

$$\frac{w(s,x)}{u(s)} = \sum_{i=1}^{\infty} \frac{\Phi_i(r_1)\Phi_i(r)}{s^2 + \omega_i^2}, \quad (2.47)$$

which is similar to equation (2.22).

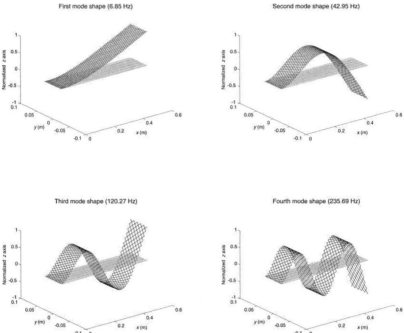

*Figure 2.12 – First mode shapes of the cantilevered aluminum beam and corresponding eigenfrequencies.*

### 2.2.3. Modal analysis of a thin plate

In the previous section the Euler-Bernoulli formulation for beams and its related modal analysis has been shown, which is valid as long as the width $b$ and thickness $h$ of the beam are much smaller than its length $L$. It is then of interest to extend this discussion to consider elastic motion in two dimensions, i.e. vibrations of thin plates, whose length and width are of similar order of magnitude and much larger than the thickness.

Let us consider a thin plate of dimensions $a \times b \times h$, as shown in Figure 2.13, and let us assume the two following conditions:

- the plate has a uniform thickness, and this is much smaller than the other two dimensions,

- the shear deformation, stress in lateral directions and rotational inertia of the plate are ignored.

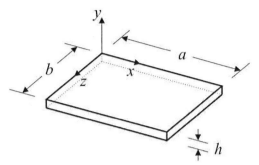

**Figure 2.13 – An elementary unconstrained thin plate.**

Let $y$ be the out-of-plane deflection, that is related to the longitudinal strain over the $x$ and $z$ directions, respectively $S_x$ and $S_z$, and to the shear strain $\gamma_{xz}$ by the following equations:

$$S_x = -y\frac{\partial^2 w}{\partial x^2},$$

$$S_z = -y\frac{\partial^2 w}{\partial z^2},$$

$$\gamma_{xz} = -2y\frac{\partial^2 w}{\partial x \partial z}. \tag{2.48}$$

From Hooke's law, strains are related to stresses as follows:

$$S_x = \frac{1}{E}(\sigma_x - \upsilon\sigma_z),$$

$$S_z = \frac{1}{E}(\sigma_z - \upsilon\sigma_x),$$

$$\gamma_{xz} = \frac{1}{G}\tau_{xz} = 2\frac{(1+\upsilon)}{E}\tau_{xz}, \tag{2.49}$$

where $\sigma_x$ and $\sigma_z$ are the longitudinal stresses in $x$ and $z$ directions, $\tau_{xz}$ the shear stress, $\upsilon$ the Poisson's ratio, $E$ is the Young's modulus of the plate material and $G$ the shear modulus. By combining equations (2.48) and (2.49), one can obtain the following expressions for the stresses:

$$\sigma_x = -\frac{Ey}{(1-\upsilon^2)}\left(\frac{\partial^2 w}{\partial x^2} + \upsilon\frac{\partial^2 w}{\partial z^2}\right),$$

$$\sigma_z = -\frac{Ey}{(1-\upsilon^2)}\left(\frac{\partial^2 w}{\partial z^2} + \upsilon\frac{\partial^2 w}{\partial x^2}\right),$$

$$\tau_{xz} = -\frac{Ey}{(1+\upsilon)\upsilon}\frac{\partial^2 w}{\partial x\partial z}. \tag{2.50}$$

The moments per length, $M_x$ and $M_z$, and the torsional moment per length, $M_{xz}$, can be calculated by integrating the corresponding stress across the thickness:

$$M_x = \int_{-\frac{h}{2}}^{\frac{h}{2}} y\sigma_x dz = -D\left(\frac{\partial^2 w}{\partial x^2} + \upsilon\frac{\partial^2 w}{\partial z^2}\right),$$

$$M_z = \int_{-\frac{h}{2}}^{\frac{h}{2}} y\sigma_z dz = -D\left(\frac{\partial^2 w}{\partial z^2} + \upsilon\frac{\partial^2 w}{\partial x^2}\right),$$

$$M_{xz} = \int_{-\frac{h}{2}}^{\frac{h}{2}} y\tau_{xz} dy = -D(1-\upsilon)\frac{\partial^2 w}{\partial x\partial z}. \tag{2.51}$$

where $D$ is the flexural rigidity of the plate defined as:

$$D = \frac{Eh^3}{12(1-\upsilon^2)}.$$  (2.52)

After some mathematical passages [RHF03], one can get the following partial differential equation (PDE) of a thin plate under transverse vibration:

$$\rho(x,z)h\frac{\partial^2 w}{\partial t^2} + D\nabla^4 w(t,x,z) = \frac{\partial^2 M_{px}}{\partial x^2} + \frac{\partial^2 M_{pz}}{\partial z^2} + p_z(t,x,z),$$  (2.53)

where $w$ is the vertical translation along the $y$-axis, $\rho(x,z)$ is the density of the plate, $dM_{px}$ and $dM_{pz}$ the external moments per length, $p_z(t,x,z)$ the pressure in $y$ direction and

$$\nabla^4 w = \frac{\partial^4 w}{\partial x^4} + 2\frac{\partial^4 w}{\partial x^2 \partial z^2} + \frac{\partial^4 w}{\partial z^4}.$$  (2.54)

As it has been shown in the previous section for the case of a thin beam, also in this case mode shapes and natural frequencies can be determined. The solution of equation (2.53) has the following form:

$$y(t,x,z) = \sum_{m=1}^{\infty}\sum_{n=1}^{\infty}\phi_{mn}(x,z)q_{mn}(t),$$  (2.55)

which is similar to equation (2.13). In order to obtain the mode shapes expression, let us consider the homogenous equation associated to equation (2.53):

$$\rho h\frac{\partial^2 y}{\partial t^2} + D\nabla^4 y(t,x,z) = 0,$$  (2.56)

where the density $\rho$ is considered constant over the plate.

By substituting equation (2.55) in equation (2.56), one can finally get the expression for the mode shapes:

$$\phi_{mn}(x, z) = A_1 \sin(\alpha x)\sin(\beta x) + A_2 \sin(\alpha x)\cos(\beta x) +$$
$$A_3 \cos(\alpha x)\sin(\beta x) + A_4 \cos(\alpha x)\cos(\beta x) +$$
$$A_5 \sinh(\overline{\alpha} x)\sinh(\overline{\beta} x) + A_6 \sinh(\overline{\alpha} x)\cosh(\overline{\beta} x) +$$
$$A_7 \cosh(\overline{\alpha} x)\sinh(\overline{\beta} x) + A_8 \cosh(\overline{\alpha} x)\cosh(\overline{\beta} x), \tag{2.57}$$

where

$$\alpha^2 + \beta^2 = \overline{\alpha}^2 + \overline{\beta}^2 = \sqrt[4]{\frac{\omega^2 \rho h}{D}}. \tag{2.58}$$

Let us consider now the case of a simply supported rectangular thin plate, that means all the edges of the plate are hinged. The boundary conditions can be written as:

$$y(t, x, z) = 0, \quad M_x = 0, \quad \forall x = 0, \quad 0 \le z \le b, \tag{2.59}$$

$$y(t, x, z) = 0, \quad M_z = 0, \quad \forall z = 0, \quad 0 \le x \le a. \tag{2.60}$$

These conditions can be expressed in terms of eigenfunctions and written as follows:

$$\phi(x, z)\big|_{x=0,a} = 0,$$

$$\phi(x, z)\big|_{z=0,b} = 0,$$

$$\ddot{\phi}(x, z)\big|_{x=0,a} = 0,$$

$$\ddot{\phi}(x, z)\big|_{z=0,b} = 0. \tag{2.61}$$

These condition will lead to $A_i = 0$, $2 \le i \le 8$. Moreover $\sin(\alpha x) = 0$ at $x = 0, a$ and $\sin(\beta z) = 0$ at $z = 0, b$. Finally, by applying the orthogonality relations among the eigenfunctions, the mode shapes can be expressed by the following equations:

$$\alpha_m a = m\pi, \quad \beta_n b = n\pi ,$$

$$\phi_{mn}(x,z) = \frac{2}{\sqrt{ab\rho h}} \sin\left(\frac{m\pi x}{a}\right) \sin\left(\frac{n\pi z}{b}\right). \tag{2.62}$$

Moreover, by coupling equations (2.57), (2.58) and (2.62), one can obtain the expression for the natural angular frequencies which is

$$\omega_{mn} = \pi^2 \sqrt{\frac{D}{\rho h}} \left(\frac{m^2}{a^2} + \frac{n^2}{b^2}\right). \tag{2.63}$$

Finally the free response of the simply supported plate is

$$y(t,x,z) = \sum_{m=1}^{\infty} \sum_{n=1}^{\infty} A_{mn}\phi_{mn}(x,z)\sin(\omega_{mn}t + \psi_{mn}) , \tag{2.64}$$

where $A_{mn}$ and $\psi_{mn}$ depend on the initial conditions.

| Parameters | Symbol | Value |
|---|---|---|
| Length | $a$ | 800 mm |
| Width | $b$ | 500 mm |
| Thickness | $h$ | 3 mm |
| Young's modulus | $E$ | 70 GPa |
| Density | $\rho$ | 2700 kg/m$^3$ |
| Poisson ratio | $\upsilon$ | 0.3 |

*Table 2.5 – Parameters of the thin simply supported rectangular aluminum plate.*

By considering a thin rectangular plate whose parameters are listed in Table 2.5, the first four mode shapes are shown in Figure 2.14.

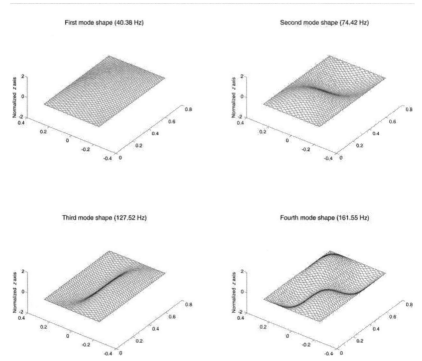

*Figure 2.14 – First mode shapes of the thin rectangular simply supported aluminum plate and corresponding eigenfrequencies.*

## 2.3.   Piezoelectric Vibrating Structures (PVS)

One way to control structural deformation or vibration is to equip the structure with elements whose actuation strain can be regulated. In particular, piezoelectric laminar translators, also called piezoelectric plates, are taken into consideration in this work. In fact, such actuators can be easily bonded on structures and in some cases they can be integrated into them, even though this case will not be discussed in this work. We call then a Piezoelectric Vibrating Structure (PVS) an elastic structure, such as a beam or a plate, with one or more piezoelectric plates used for actuating and/or sensing.

After a brief introduction on the static actuation properties of a piezoelectric material bonded on an elastic structure, it is of major interest to derive a mathematical model to describe the dynamic behaviour of the coupling between a structure and a piezoelectric material, since it will provide the basic mathematical framework for describing self sensing techniques. For this purpose we shall follow the work from Hagood [HCF90] and Moheimani [RHF03] who have derived similar models by using different approaches. After showing these two formulations, a comparison between them will be made. Finally, the problem of the optimal placement of actuators and sensors on a vibrating structure will be shortly discussed.

## 2.3.1. Piezoelectric static actuation in a PVS

In this section we consider the actuation induced by a piezoelectric plate bonded on an elastic structure, such as a beam or a plate. Considering the main directions of motions which are of interest in this work, they can be called respectively 1-D and 2-D analysis. In both cases the actuator induces surface strain to the structure through the $d_{31}$ and $d_{32}$ piezoelectric actuator mode when a voltage is applied. We also assume that $d_{31} = d_{32}$.

*1-D analysis: piezoelectric actuation of a beam*

For the 1-D analysis we shall follow the works of Crawley and Anderson [And86], [CA90]. Two mathematical models are then considered at the aim of comparison, in particular the *uniform strain* model, which considers the strain uniform in the piezoelectric actuators, and the *Bernoulli-Euler* model which assumes the strain to change linearly over the out of plane axis both in the structure and in the actuators. Moreover, both extension and bending will be considered. In Figure 2.15 the induced strain distributions in both motions and both models are depicted. We also assume a perfect bending of the actuator on the structure. In reality a thin layer of glue is usually used for bonding, which limits the actuating/sensing capabilities of the piezoelectric plates by introducing shear stresses between the piezoelectric plate and the structure. Nevertheless such case is not analyzed in this work, since it can be neglected as long as the layer of glue is uniformly distributed and much thinner than the piezoelectric actuator.

When a voltage is applied on the bonded piezoelectric actuators they will attempt to expand but will be constrained by the stiffness of the structure. This results in an extension and bending of the beam. In particular, the forces and moments are transmitted to the structure at the edges of the piezoelectric actuators. The strain distribution of the actuators and of the structure are then reported per each case in Figure 2.15.

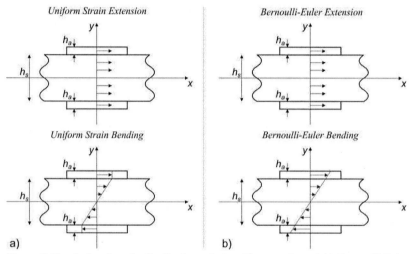

**Figure 2.15 – Induced strain distributions of a) uniform strain and b) Bernoulli-Euler extension and bending.**

*Uniform strain – Extension*

In the uniform strain model the assumption is that the strain of the piezoelectric plate actuators equals the surface strain of the structure where the piezoelectric plate is bonded. Such assumption holds when the structure thickness $h_s$ is much bigger than the piezoelectric plate actuator thickness $h_a$. In the case of uniform strain extension the strain of the structure $S_s$ equals the strain of the actuators $S_a$, and can be expressed as:

$$S_s = S_a = \frac{2\Lambda}{2 + \Psi_e},$$
(2.65)

where

$$\Psi_e = \frac{E_s A_s}{E_a A_a},\tag{2.66}$$

$E_s$ and $E_a$ are respectively the Young's modulus of the structure and of the actuators, $A_s$ and $A_a$ their cross section areas, and $\Lambda$ is the actuation strain caused by piezoelectricity or any other phenomenon (thermoelasticity, electrostriction…). In particular, for purely piezoelectric actuation, the actuation strain can be expressed as:

$$\Lambda = d_{31} \frac{V}{h_a}.\tag{2.67}$$

*Uniform strain – Bending*

In this case the strain of the actuators is uniform and equals the surface strain of the structure where the actuators are bonded.

The upper actuator strain can be expressed as:

$$S_s^{surf} = S_s\big|_{y=\frac{h_s}{2}} = S_a = \frac{6\Lambda}{6 + \Psi_b},\tag{2.68}$$

where

$$\Psi_b = \frac{12 E_s I_s}{h_s^2 E_a A_a}\tag{2.69}$$

is a measure of the relative stiffnesses of the structure and one actuator. Finally, for rectangular cross-section of the beam, one can get that

$$I_s = \frac{A_s h_s^2}{12},\tag{2.70}$$

and consequently the relative stiffness parameter $\Psi_b$ becomes:

$$\Psi_b = \frac{E_s A_s}{E_a A_a} .$$ (2.71)

The lower actuator induced strain instead is:

$$S_s^{surf} = S_s\big|_{y=-\frac{h_s}{2}} = -\frac{6\Lambda}{6+\Psi_b} .$$ (2.72)

The curvature induced in the structure is:

$$\kappa = -\frac{2}{h_s}\frac{6\Lambda}{6+\Psi_b} .$$ (2.73)

A plot of the normalized induced curvature $\kappa h_s / 2\Lambda$ is shown in Figure 2.16 at different Young's modulus ratio $E_s / E_a$ as a function of the thickness ratio $H = h_s / h_a$, and assuming the actuators and the beam to have same width.

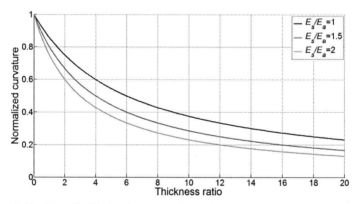

*Figure 2.16 – Normalized induced curvature for uniform strain bending model.*

Figure 2.16 shows how much authority a piezoelectric plate has on a structure according to the thickness ratio $H$ and the Young's modulus ratio according to the uniform strain model. In particular the maximum authority corresponds to low structure-to-actuator thicknesses. This does not meet confirmation in experimental observations.

*Bernoulli-Euler – Extension*

Considering Figure 2.15, it is assumed in this model that both structure and actuators strain are uniform for extension (as in the uniform strain model) and linear with $z$ for bending. Consequently, the induced extension strain has the same expression as for the uniform strain model:

$$S_s = S_a = \frac{2\Lambda}{2+\Psi_e} , \tag{2.74}$$

where

$$\Psi_e = \frac{E_s A_s}{E_a A_a} . \tag{2.75}$$

*Bernoulli-Euler – Bending*

In this formulation both the actuator and the structure undergo the same strain, so we refer to a unique induced bending strain $S$. Let us assume that the distance between the inner face of the actuator and the neutral axis is $d = h_s / 2$. The induced bending strain $S$ according to the Bernoulli-Euler model is:

$$S = -\kappa y = \frac{2E_a A_a \left( \dfrac{h_s}{2} + \dfrac{h_a}{2} \right)\Lambda}{E_s \left( \dfrac{b_s h_s^3}{12} \right) + 2E_a I_a} y , \tag{2.76}$$

where $b_s$ is the width of the structure and

$$I_a = A_a \left( \frac{h_s^2}{4} + \frac{h_s h_a}{2} + \frac{h_a^2}{3} \right), \tag{2.77}$$

which is the moment of inertia per each actuator. Consequently, by putting equation (2.77) into equation (2.76), the induced bending strain can be written as:

$$S = -\kappa y = \frac{12\left(h_s + h_a\right)\Lambda}{\left(\dfrac{E_s A_s}{E_a A_a} + 6\right)h_s^2 + 12 h_s h_a + 8 h_a^2}\, y\,.$$

(2.78)

By using the stiffness ratio $\Psi$ (equation (2.69)) and the thickness ratio $H = h_s / h_a$, for a rectangular cross-section beam ($\Psi_b = \Psi_e$), one can express the induced bending strain as:

$$S = -\kappa y = \frac{6\left(1 + \dfrac{1}{H}\right)\Lambda\left(\dfrac{2}{h_s}\right)}{\left(6 + \Psi\right) + \dfrac{12}{H} + \dfrac{8}{H^2}}\, y\,.$$

(2.79)

It is now of interest to analyze the normalized induced curvature $\kappa h_s / 2\Lambda$ in the case of Bernoulli-Euler formulation shown in Figure 2.17 and compare it to the one shown in Figure 2.16.

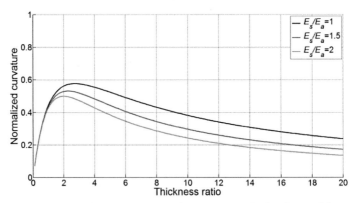

Figure 2.17 – Normalized induced curvature for Bernoulli-Euler bending model.

Considered that the Bernoulli-Euler model is more reliable than the uniform strain one and that is confirmed also by FEM simulations, one can observe that the authority of a piezoelectric actuator over a beam is maximed for a thickness ratio between 2 and 3, depending on the Young's modulus ratios chosen in Figure 2.17. In fact, if a piezoelectric material is much thicker than the structure ($h_a \ll h_s$), the induced bending strain is as low as the case of a piezoelectric material much thinner than the structure ($h_a \gg h_s$). Thus it is very important to choose a correct thickness ratio when dimensioning the PVS, and this result will be considered again in section 3.2 since it is an important parameter for dimensioning self sensing actuators.

### 2-D analysis: piezoelectric actuation of an isotropic plate

In this section extension and bending strains are considered in the case of two-dimensional plate structures. In particular, the uniform strain model of the previous section will be extended to this case and a model of induced strain actuation is derived using classical laminated plate theory. In both cases, the attention will be focused only on isotropic plates with actuation symmetric to the neutral axes. For more general solutions one can refer to [TW59], [Lei69]. Moreover only the main results will be shown and shortly commented, since the mathematical derivation is not of interest at the aim of this work.

*Uniform strain – Extension*

The induced strain extension along the $x$ and $z$ axes is expressed as:

$$S_x = S_z = \frac{2}{2 + \dfrac{E_s h_s}{E_a h_a}} \Lambda, \qquad \gamma_{xz} = 0, \qquad (2.80)$$

where $\gamma_{xz}$ is the shear strain.

This result is identical to the case of a beam with equal width structure and actuators shown in equation (2.65).

*Uniform strain – Bending*

For the case of induced bending strain, the resulting strains of the structure and of the actuators can be expressed as:

$$S_{x_s} = S_{z_s} = \kappa_s = \frac{6}{6 + \dfrac{E_s h_s}{E_a h_a}} \Lambda\left(\frac{2}{h_s}\right) y, \qquad \gamma_{xz_s} = y\kappa_{xz_s} = 0, \qquad (2.81)$$

$$S_{x_a} = S_{z_a} = \kappa_s \left(\frac{h_s}{2}\right) = \frac{6}{6 + \dfrac{E_s h_s}{E_a h_a}} \Lambda y, \qquad \gamma_{xz_a} = \frac{h_s}{2}\kappa_{xz_a} = 0. \qquad (2.82)$$

which is again the identical expression for the case of a beam with equal width structure and actuators shown in equation (2.72).

*Laminated Plate – Extension*

In the case of the laminated plate theory, the induced extension strain can be expressed as:

$$S_x = S_z = \frac{2}{2 + \dfrac{E_s h_s}{E_a h_a}} \Lambda, \qquad \gamma_{xz} = 0, \qquad (2.83)$$

which is identical to equation (2.80).

*Laminated Plate – Bending*

The expression for induced bending strain under laminated plate theory is:

$$\kappa_x = \kappa_z = \frac{2E_a h_a\left(d + \dfrac{h_a}{2}\right)}{\left\{ E_s\left[\dfrac{h_s^3}{12} - h_a\left(d^2 + dh_a + \dfrac{h_a^3}{3}\right)\right] + 2E_a h_a\left(d^2 + dh_a + \dfrac{h_a^3}{3}\right)\right\}} \Lambda,$$

$$\kappa_{xz} = 0, \tag{2.84}$$

where $d$ is the distance between the inner face of the actuator and the neutral axes. Assuming $d = h_s / 2$, one can write that:

$$S_x = S_z = -\kappa_x y = -\kappa_z y = \frac{6\left(1 + \dfrac{1}{H}\right)\Lambda\left(\dfrac{2}{h_s}\right)}{\left(6 + \Psi\right) + \dfrac{12}{H} + \dfrac{8}{H^2}} y, \qquad \kappa_{xz} = 0, \tag{2.85}$$

where $\Psi$ and $H$ are defined in the previous section. The right side of equation (2.85) is identical to equation (2.79). The laminated plate model can be considered as a generalization of the one-dimensional Euler-Bernoulli model shown in the previous section. In fact, for isotropic and symmetric structure and actuators the solutions are identical.

Such results are an important contribution to the dimensioning of self sensing actuators, as it will be discussed in section 3.2.

## 2.3.2.  Dynamic model of a PVS – Rayleigh-Ritz formulation

In this section a Rayleigh-Ritz formulation is used in order to derive the equations of motion of a PVS. In particular, the generalized form of the Hamilton's principle for a general electromechanical system is provided from [CKK68] and is:

$$\int_{t_1}^{t_2} \left[ \partial \left( K - U + W_e - W_m \right) + \partial W \right] dt = 0, \tag{2.86}$$

where $K$ and $U$ are respectively kinetic and potential energies, $W_e$ and $W_m$ the electrical and magnetic energies, and $W$ is the work of nonconservative forces. In the case of piezoelectric materials the magnetic contribution $W_m$ is neglected. Moreover, considered also that only discrete applied external forces are applied at the point locations $x_i$ and that charges are applied at a discrete set of piezoelectric electrodes, such quantities are defined as:

$$K = \int_{V_s} \frac{1}{2} \rho_s \dot{\mathbf{u}}^T \dot{\mathbf{u}} + \int_{V_p} \frac{1}{2} \rho_p \dot{\mathbf{u}}^T \dot{\mathbf{u}} , \qquad (2.87)$$

$$U = \int_{V_s} \frac{1}{2} \mathbf{S}^T \boldsymbol{\sigma} + \int_{V_p} \frac{1}{2} \mathbf{S}^T \boldsymbol{\sigma} , \qquad (2.88)$$

$$W_e = \int_{V_p} \frac{1}{2} \mathbf{E}^T \mathbf{D} , \qquad (2.89)$$

$$\partial W = \sum_{i=1}^{nf} \partial \mathbf{u}(x_i) \mathbf{f}(x_i) - \sum_{j=1}^{nq} \partial \varphi_j q_j , \qquad (2.90)$$

where the quantities definitions are shown in Table 2.6 and the apex $T$ refers to the transposition operator since temperature is not considered in this section.

Moreover in equation (2.90) $nf$ and $nq$ are respectively the number of forces acting on the PVS and the number of applied charges.

The vectors used are written in terms of their components as:

$$\mathbf{D} = \begin{bmatrix} D_1 \\ D_2 \\ D_3 \end{bmatrix}, \quad \mathbf{E} = \begin{bmatrix} E_1 \\ E_2 \\ E_3 \end{bmatrix}, \quad \mathbf{S} = \begin{bmatrix} S_{11} \\ S_{22} \\ S_{33} \\ 2S_{23} \\ 2S_{13} \\ 2S_{12} \end{bmatrix} = \begin{bmatrix} S_1 \\ S_2 \\ S_3 \\ S_4 \\ S_5 \\ S_6 \end{bmatrix}, \quad \boldsymbol{\sigma} = \begin{bmatrix} \sigma_{11} \\ \sigma_{22} \\ \sigma_{33} \\ \sigma_{23} \\ \sigma_{13} \\ \sigma_{12} \end{bmatrix} = \begin{bmatrix} \sigma_1 \\ \sigma_2 \\ \sigma_3 \\ \sigma_4 \\ \sigma_5 \\ \sigma_6 \end{bmatrix} . \qquad (2.91)$$

It is now appropriate to introduce the constitutive relations, both for the structure and for the piezoelectric material. The structure constitutive equations expressed in terms of global coordinates are expressed by:

$$\boldsymbol{\sigma} = \mathbf{c}_s \mathbf{S}, \tag{2.92}$$

that is the generalized Hooke's law, where $\mathbf{c}_s$ is the elasticity matrix.

By recalling the piezoelectric constitutive equations (2.4), and by considering that the $3^{\text{rd}}$ direction is associated with the direction of polarization, the constitutive equations of a piezoelectric material written in a matricial form are:

$$\begin{bmatrix} \mathbf{D}' \\ \boldsymbol{\sigma}' \end{bmatrix} = \begin{bmatrix} \boldsymbol{\varepsilon}^S & \mathbf{e} \\ -\mathbf{e}^T & \mathbf{c}^E \end{bmatrix} \begin{bmatrix} \mathbf{E}' \\ \mathbf{S}' \end{bmatrix}, \tag{2.93}$$

| Symbol | Physical Quantity |
|--------|-------------------|
| $\mathbf{D}(x)$ | vector of electrical displacement |
| $\mathbf{E}(x)$ | vector of the electrical field in the material |
| $\mathbf{S}(x)$ | vector of material strain |
| $\boldsymbol{\sigma}(x)$ | vector of material stress |
| $\varphi(x)$ | scalar of electrical potential |
| $\mathbf{u}(x_i)$ | vector of mechanical displacement |
| $\mathbf{f}(x_i)$ | vector of applied force at location $x_i$ |
| $q_j$ | charge applied at electrode $j$ |

*Table 2.6 – Definition of the physical quantities for the Rayleigh-Ritz formulation.*

where ( )' denotes variables referring to the local coordinate frame attached to the polarization direction. In order to relate the constitutive equation of (2.92) to the global coordinate system, we define the vector $\mathbf{p}(\mathbf{x})$ representing the direction of piezoelectric poling, and consequently we define the rotation matrices $\mathbf{R}_S(\mathbf{x},\mathbf{p})$ and $\mathbf{R}_E(\mathbf{x},\mathbf{p})$ relating the local coordinate system to the global one:

$$\mathbf{S}' = \mathbf{R}_S(\mathbf{x},\mathbf{p})\mathbf{S} \text{ and } \mathbf{E}' = \mathbf{R}_E(\mathbf{x},\mathbf{p})\mathbf{E}. \tag{2.94}$$

The structural rotation matrix $\mathbf{R}_S(\mathbf{x},\mathbf{p})$ is commonly used in elasticity, while $\mathbf{R}_E(\mathbf{x},\mathbf{p})$ is a matrix of direction cosines. Finally the constitutive equations of the piezoelectric material referring to the global coordinate system are:

$$\begin{bmatrix} \mathbf{D} \\ \boldsymbol{\sigma} \end{bmatrix} = \begin{bmatrix} \mathbf{R}_E^T \boldsymbol{\varepsilon}^S \mathbf{R}_E & \mathbf{R}_E^T \mathbf{e} \mathbf{R}_S \\ -\mathbf{R}_S^T \mathbf{e}^T \mathbf{R}_E & \mathbf{R}_S^T \mathbf{c}^E \mathbf{R}_S \end{bmatrix} \begin{bmatrix} \mathbf{E} \\ \mathbf{S} \end{bmatrix}. \tag{2.95}$$

After defining the constitutive equations both for the structure and the piezoelectric materials, the strain-displacement and field-potential relations are:

$$\mathbf{S} = \mathbf{L}_u \mathbf{u}(\mathbf{x}) \quad \text{and} \quad \mathbf{E} = \mathbf{L}_\varphi \varphi(\mathbf{x}) = -\nabla \varphi(\mathbf{x}), \tag{2.96}$$

where $\mathbf{L}_u$ is the linear differential operator for the particular elasticity problem and $\mathbf{L}_\varphi$ is the gradient operator.

The displacement and potential function can be expressed in terms of generalized coordinates according to the Rayleigh-Ritz formulation:

$$\mathbf{u}(\mathbf{x},t) = \boldsymbol{\Psi}_r(\mathbf{x})\mathbf{r}(t) = \begin{bmatrix} \boldsymbol{\Psi}_{r_1}(\mathbf{x}) & \cdots & \boldsymbol{\Psi}_{r_n}(\mathbf{x}) \end{bmatrix} \begin{bmatrix} r_1(t) \\ \vdots \\ r_n(t) \end{bmatrix}, \tag{2.97}$$

$$\varphi(\mathbf{x},t) = \boldsymbol{\Psi}_v(\mathbf{x})\mathbf{v}(t) = \begin{bmatrix} \boldsymbol{\Psi}_{v_1}(\mathbf{x}) & \cdots & \boldsymbol{\Psi}_{v_m}(\mathbf{x}) \end{bmatrix} \begin{bmatrix} v_1(t) \\ \vdots \\ r_m(t) \end{bmatrix}, \tag{2.98}$$

where $r_i$ is the generalized mechanical coordinate and $v_i$ the generalized electrical coordinate. In particular, when the piezoelectric actuators are driven by a voltage amplifier, it is convenient to assume that the $v_i$ coordinates represent the physical voltage applied at the piezoelectric electrodes. Moreover, $\boldsymbol{\Psi}_{r_i}$ and $\boldsymbol{\Psi}_{v_i}$ are respectively the assumed displacement and potential distribution, and they need to be differentiable to the orders of $\mathbf{L}_u$ and $\mathbf{L}_\varphi$. In order to simplify the following equations, we introduce the field basis functions:

$$\mathbf{S}(\mathbf{x},t) = N_r(\mathbf{x})\mathbf{r}(t) \text{ and } \mathbf{E}(\mathbf{x},t) = N_v(\mathbf{x})\mathbf{v}(t), \tag{2.99}$$

where

$$N_r(\mathbf{x}) = \mathbf{L}_u\mathbf{\Psi}_r(x) \text{ and } N_v(\mathbf{x}) = \mathbf{L}_\varphi\mathbf{\Psi}_v(x). \tag{2.100}$$

Finally, we can derive the equations of motion of the PVS by substituting the needed quantities into equation (2.86):

$$\int_{t_1}^{t_2} [\int_{V_s} \rho_s \delta\ddot{\mathbf{u}}^T \dot{\mathbf{u}} + \int_{V_p} \rho_p \delta\ddot{\mathbf{u}}^T \dot{\mathbf{u}} - \int_{V_s} \delta\mathbf{S}^T \mathbf{c}_s \mathbf{S} - \int_{V_p} \delta\mathbf{S}^T \mathbf{R}_S^T \mathbf{c}^E \mathbf{R}_S \mathbf{S}$$
$$+ \int_{V_p} \delta\mathbf{S}^T \mathbf{R}_S^T \mathbf{e}^T \mathbf{R}_E \mathbf{E} + \int_{V_p} \delta\mathbf{E}^T \mathbf{R}_E^T \mathbf{e} \mathbf{R}_s \mathbf{S} + \int_{V_p} \delta\mathbf{E}^T \mathbf{R}_E^T \boldsymbol{\varepsilon}^S \mathbf{R}_E \mathbf{E} \tag{2.101}$$
$$+ \sum_{i=1}^{nf} \partial\mathbf{u}(x_i)\mathbf{f}(x_i) - \sum_{j=1}^{nq} \partial\varphi_j q_j] dt = 0.$$

From such equation we can get the following actuator and sensor equations for a PVS:

$$\left(\mathbf{M}_s + \mathbf{M}_p\right)\ddot{\mathbf{r}} + \left(\mathbf{K}_s + \mathbf{K}_p\right)\mathbf{r} - \mathbf{\Theta}\mathbf{v} = \mathbf{B}_f\mathbf{f}$$
$$\mathbf{\Theta}^T\mathbf{r} + \mathbf{C}_p\mathbf{v} = \mathbf{B}_q\mathbf{q}, \tag{2.102}$$

where the mass matrixes are defined as

$$\mathbf{M}_s = \int_{V_s} \mathbf{\Psi}_r^T(\mathbf{x})\rho_s(\mathbf{x})\mathbf{\Psi}_r(\mathbf{x}) \qquad \mathbf{M}_p = \int_{V_p} \mathbf{\Psi}_r^T(\mathbf{x})\rho_p(\mathbf{x})\mathbf{\Psi}_r(\mathbf{x}), \tag{2.103}$$

the stiffness matrixes are

$$\mathbf{K}_s = \int_{V_s} N_r^T(\mathbf{x})\mathbf{c}_s N_r(\mathbf{x}) \qquad \mathbf{K}_p = \int_{V_p} N_r^T(\mathbf{x})\mathbf{R}_s^T \mathbf{c}^E \mathbf{R}_s N_r(\mathbf{x}), \tag{2.104}$$

the piezoelectric capacitance matrix and the electromechanical coupling matrix are

$$\mathbf{C}_p = \int_{V_p} N_v^T\left(\mathbf{x}\right)\mathbf{R}_E^T\boldsymbol{\varepsilon}^S\mathbf{R}_E N_v\left(\mathbf{x}\right) \quad \boldsymbol{\Theta} = \int_{V_p} N_r^T\left(\mathbf{x}\right)\mathbf{R}_s^T\mathbf{e}^T\mathbf{R}_E N_v\left(\mathbf{x}\right), \quad (2.105)$$

and finally the forcing matrixes are defined as

$$\mathbf{B}_f = \begin{bmatrix} \Psi_{r_1}^T\left(x_{f_1}\right) & \cdots & \Psi_{r_1}^T\left(x_{f_{nf}}\right) \\ \vdots & & \vdots \\ \Psi_{r_n}^T\left(x_{f_1}\right) & \cdots & \Psi_{r_n}^T\left(x_{f_{nf}}\right) \end{bmatrix}, \qquad (2.106)$$

$$\mathbf{B}_q = \begin{bmatrix} \Psi_{v_1}^T\left(x_{q_1}\right) & \cdots & \Psi_{v_1}^T\left(x_{q_{nq}}\right) \\ \vdots & & \vdots \\ \Psi_{v_m}^T\left(x_{q_1}\right) & \cdots & \Psi_{v_m}^T\left(x_{q_{nq}}\right) \end{bmatrix}, \qquad (2.107)$$

where $n$ and $m$ are respectively the number of mechanical and electrical degrees of freedom.

The physical voltages applied at the piezoelectric materials $\mathbf{v}_p$ are related to the generalized voltage coordinates $\mathbf{v}$ through the matrix $\mathbf{B}_q$ :

$$\mathbf{v}_p = \mathbf{B}_q\mathbf{v} . \qquad (2.108)$$

Equations (2.102) describe the motion of a PVS with a certain number of piezoelectric materials and arbitrary electrode arrangement and piezoelectric geometry. In particular, the actuator matricial differential equation is expressed in terms of the mechanical generalized coordinates $r$, whose forcing quantities are the voltages $v$ and forces $f$. The sensor equation is then useful to calculate the amount of charge flowing through the piezoelectric materials.

By taking the Laplace transform and grouping the mechanical and electrical impedances, equations (2.102) can be written as:

$$\left(\frac{1}{sZ_m(s)}\right)\mathbf{r} - \mathbf{\Theta v} = \mathbf{B}_f\mathbf{f}$$
$$\left(\frac{1}{sZ_e(s)}\right)\mathbf{v} + \mathbf{\Theta}^T\mathbf{r} = \mathbf{B}_q\mathbf{q}$$
(2.109)

where:

$$\mathbf{Z}_m(s) = \left(\mathbf{M}_s + \mathbf{M}_p\right)s + \left(\mathbf{K}_s + \mathbf{K}_p\right)/s, \qquad \mathbf{Z}_e(s) = 1/s\mathbf{C}_p,$$
(2.110)

are respectively the mechanical and electrical impedance. This formulation shows the symmetry between the mechanical and electrical behaviour in a PVS.

It is of interest to obtain the state space equations of a PVS with voltage driven electrodes. Assuming that the generalized voltage coordinates $\mathbf{v}$ are the physical voltages $\mathbf{v}_p$, then $\mathbf{B}_q = \mathbf{I}$, which is the identity matrix. The state space equations are written as follows:

$$\dot{\xi} = \mathbf{A}\xi + \mathbf{B}\tau$$
$$\chi = \mathbf{C}\xi + \mathbf{D}\tau$$
(2.111)

where the state vector $\xi$, the input vector $\tau$, the state matrix $\mathbf{A}$ and the forcing matrix $\mathbf{B}$ are defined as follows:

$$\xi = \begin{bmatrix} \mathbf{r} \\ \dot{\mathbf{r}} \end{bmatrix}, \qquad \tau = \begin{bmatrix} \mathbf{f} \\ \mathbf{v} \end{bmatrix},$$
(2.112)

$$\mathbf{A} = \begin{bmatrix} 0 & \mathbf{I} \\ -\mathbf{M}^{-1}\mathbf{K} & -\mathbf{M}^{-1}\mathbf{\Delta} \end{bmatrix}, \qquad \mathbf{B} = \begin{bmatrix} 0 & 0 \\ -\mathbf{M}^{-1}\mathbf{B}_f & \mathbf{M}^{-1}\mathbf{\Theta} \end{bmatrix},$$
(2.113)

$$\mathbf{M} = \mathbf{M}_s + \mathbf{M}_p, \qquad \mathbf{K} = \mathbf{K}_s + \mathbf{K}_p,$$
(2.114)

where $\mathbf{\Delta}$ is an arbitrary damping matrix. Moreover the output equation matrixes $\mathbf{C}$ and $\mathbf{D}$ are dependent on the observed outputs $\chi$ and vary according to the needs of modeling.

Let us now apply this formulation to two cases: a cantilevered piezoelectric beam and a simply supported piezoelectric plate.

***Example 1: a cantilevered piezoelectric beam model***

In this example we consider the case of a cantilevered beam with length $L$ actuated by a piezoelectric plate, whose end points coordinates are $a_1$ and $a_2$ as shown in Figure 2.18. In order to obtain a state space dynamic model, the first step is to define the rotation matrixes $\mathbf{R}_S(\mathbf{x},\mathbf{p})$ and $\mathbf{R}_E(\mathbf{x},\mathbf{p})$ to pass from the local piezoelectric reference system $(x',y',z')$ to the structure reference system $(x,y,z)$. Considered the polarization direction in the negative $y$ direction, these matrixes can be defined as:

$$\mathbf{R}_S = \begin{bmatrix} 1 & 0 & 0 & 0 & 0 & 0 \\ 0 & 0 & 1 & 0 & 0 & 0 \\ 0 & 1 & 0 & 0 & 0 & 0 \\ 0 & 0 & 0 & 0 & -1 & 0 \\ 0 & 0 & 0 & -1 & 0 & 0 \\ 0 & 0 & 0 & 0 & 0 & 1 \end{bmatrix}, \tag{2.115}$$

$$\mathbf{R}_E = \begin{bmatrix} 1 & 0 & 0 \\ 0 & 0 & 1 \\ 0 & -1 & 0 \end{bmatrix}. \tag{2.116}$$

Afterwards, by considering the Euler-Bernoulli assumption for a beam and constant electric field strength through the thickness of the piezoelectric plate, the strain-displacement operator $\mathbf{L}_u$ and the field-potential operator $\mathbf{L}_\varphi$ are written as:

$$\mathbf{L}_u = \begin{bmatrix} 0 & -y\dfrac{\partial^2}{\partial x^2} & 0 \\[2mm] 0 & \upsilon y\dfrac{\partial^2}{\partial x^2} & 0 \\[2mm] 0 & 0 & 0 \\[1mm] 0 & 0 & 0 \\[1mm] 0 & 0 & 0 \\[1mm] 0 & 0 & 0 \end{bmatrix}, \qquad (2.117)$$

$$\mathbf{L}_\varphi = \begin{bmatrix} 0 \\[1mm] -\dfrac{\delta}{\delta y} \\[2mm] 0 \end{bmatrix}, \qquad (2.118)$$

where $\upsilon$ is the Poisson's ratio of the beam.

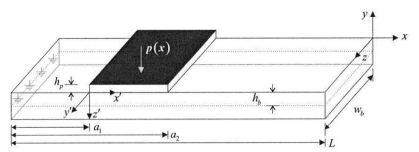

*Figure 2.18 – A simple cantilevered beam with a bonded piezoelectric plate.*

At this point, the shapes of the displacement field and electrical potential have to be decided. In particular, regarding to the displacement field, we consider the first number $N$ modal shapes plus a static mode that represents the bending of the beam when a voltage is applied to the piezoelectric plate.

This static mode can be modeled as follows:

$$\Phi^{st}(x) = \begin{cases} 0 & x < a_1 \\ \left(\dfrac{x - a_1}{a_2 - a_1}\right)^2 & a_1 \le x \le a_2 \\ 1 + 2\dfrac{x - a_2}{a_2 - a_1} & x \ge a_2 \end{cases} \tag{2.119}$$

Finally, let us consider for example the first five resonant modes ($N = 5$), the displacement vector can be written as:

$$\mathbf{u} = \begin{bmatrix} 0 & 0 & 0 & 0 & 0 & 0 \\ \Phi_1 & \Phi_2 & \Phi_3 & \Phi_4 & \Phi_5 & \Phi^{st} \\ 0 & 0 & 0 & 0 & 0 & 0 \end{bmatrix} \begin{bmatrix} r_1 \\ r_2 \\ r_3 \\ r_4 \\ r_5 \\ r_6 \end{bmatrix} = \mathbf{\Psi}_r \mathbf{r}, \tag{2.120}$$

where the mode shapes $\Phi_i$ for a cantilevered beam are shown in equation (2.45). Similarly the potential field within the piezoelectric plate is expressed in terms of the voltage at its electrodes $v_p$:

$$\varphi = \mathbf{\Psi}_v v, \tag{2.121}$$

where

$$\mathbf{\Psi}_v(x) = \begin{cases} 0 & x < a_1 \\ \dfrac{y - h_b}{h_p} & a_1 \le x \le a_2, y > 0, \\ 0 & x \ge a_2 \end{cases} \tag{2.122}$$

where $h_p$ is the piezoelectric thickness, while $h_b$ is half of the thickness of the beam.

In order to simplify the practical implementation of this model for computer applications, a matrix **RD** is defined as:

$$\mathbf{RD} = \begin{bmatrix} 0 & -1 & 0 \\ 0 & \upsilon & 0 \\ 0 & 0 & 0 \\ 0 & 0 & 0 \\ 0 & 0 & 0 \\ 0 & 0 & 0 \end{bmatrix}, \tag{2.123}$$

so that the strain displacement operator can be written, according to equation (2.100), as:

$$N_r(\mathbf{x}) = \mathbf{L}_u \mathbf{\Psi}_r(\mathbf{x}) = y\mathbf{RD}\mathbf{\Psi}''_r(\mathbf{x}), \tag{2.124}$$

where $\mathbf{\Psi}''_r(\mathbf{x}) = \dfrac{\partial^2 \mathbf{\Psi}_r(\mathbf{x})}{\partial x^2}$. Finally, by substituting the obtained expressions in equations (2.103)-(2.107), one can get:

$$\mathbf{M}_s = 2h_b \rho_s w_b \int_0^L \mathbf{\Psi}_r^T(x) \mathbf{\Psi}_r(x) \mathrm{d}x, \tag{2.125}$$

$$\mathbf{M}_p = h_p \rho_p w_p \int_{a_1}^{a_2} \mathbf{\Psi}_r^T(x) \mathbf{\Psi}_r(x) \mathrm{d}x, \tag{2.126}$$

$$\mathbf{K}_s = 2\frac{h_b^3}{3} w_b \int_0^L \mathbf{\Psi}_r''^T(x) \mathbf{RD}^T \mathbf{c}_s \mathbf{RD}\mathbf{\Psi}_r''(x) \mathrm{d}x, \tag{2.127}$$

$$\mathbf{K}_p = \left[ \frac{(h_b + h_p)^3}{3} - \frac{h_b^3}{3} \right] w_b \int_{a_1}^{a_2} \mathbf{\Psi}_r''^T(x) \mathbf{RD}^T \mathbf{R}_s^T \mathbf{c}^E \mathbf{R}_s \mathbf{RD}\mathbf{\Psi}_r''(x) \mathrm{d}x, \tag{2.128}$$

$$\mathbf{C}_p = N_v^T(x) \mathbf{R}_E^T \mathbf{\varepsilon}^S \mathbf{R}_E N_v(x) \left[ h_p (a_2 - a_1) w_b \right], \tag{2.129}$$

$$\Theta = \left[ \frac{\left(h_b + h_p\right)^2}{2} - \frac{h_b^2}{2} \right] w_b \int_{a_1}^{a_2} \Psi_r''^T (x) \mathbf{R} \mathbf{D}^T \mathbf{R}_s^T \mathbf{e}^T \mathbf{R}_E N_v (x) dx , \qquad (2.130)$$

where $\rho_s$ and $\rho_p$ are the beam and piezoelectric plate densities and $w_b$ the width of the beam (we assume the piezoelectric material to have the same width of the beam). Moreover the matrix $\Delta$ used in the matrix $\mathbf{A}$ of equation (2.113) can be expressed as:

$$\Delta = d\mathbf{I} , \qquad (2.131)$$

where $d$ is an arbitrary scalar damping factor. A usual choice for aluminum structures is $d = 0.01$.

The state space model formulation then is directly obtained by using the same formulation shown in equations (2.111)-(2.114).

### *Example 2: a simply-supported piezoelectric plate model*

A similar procedure as the one just shown can be applied to the case of a simply supported rectangular plate of dimensions $L_1$ times $L_2$ actuated by a piezoelectric rectangular plate whose ends on the $x$ and $y$ axis are respectively $a_1$ and $a_2$, $b_1$ and $b_2$, as shown in Figure 2.19.

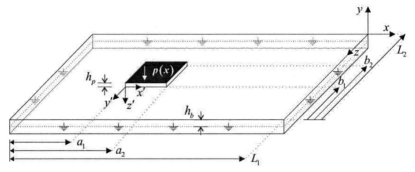

*Figure 2.19 – A simply supported plate with a bonded piezoelectric plate.*

The rotation matrixes $\mathbf{R}_S(\mathbf{x},\mathbf{p})$ and $\mathbf{R}_E(\mathbf{x},\mathbf{p})$ are defined in the same way as in equations (2.115) and (2.116), as well as the field-potential operator $\mathbf{L}_\varphi$ which is expressed in equation (2.118). The strain-displacement operator $\mathbf{L}_u$ has to be instead redefined as:

$$
\mathbf{L}_u = \begin{bmatrix}
0 & -y\dfrac{\partial^2}{\partial x^2} + \upsilon y\dfrac{\partial^2}{\partial z^2} & 0 \\[2mm]
0 & \upsilon y\dfrac{\partial^2}{\partial x^2} - y\dfrac{\partial^2}{\partial z^2} & 0 \\[2mm]
0 & 0 & 0 \\
0 & 0 & 0 \\
0 & 0 & 0 \\
0 & 0 & 0
\end{bmatrix},
\tag{2.132}
$$

where $\upsilon$ is the Poisson's ratio of the plate.

The shapes of the displacement field are expressed in equations (2.62), while the static mode shapes can be defined in this case as:

$$
\Phi^{st}(x,z) = \sin\frac{\pi(x-a_1)}{a_2-a_1}\sin\frac{\pi(z-b_1)}{b_2-b_1}.
\tag{2.133}
$$

Considering the first 5 modes of the plate, the matrix $\mathbf{\Psi}_r(x,z)$ and its two derivates respect to $x$ and $z$ can then be written as:

$$
\mathbf{\Psi}_r = \begin{bmatrix}
0 & 0 & 0 & 0 & 0 & 0 \\
\Phi_1 & \Phi_2 & \Phi_3 & \Phi_4 & \Phi_5 & \Phi^{st} \\
0 & 0 & 0 & 0 & 0 & 0
\end{bmatrix},
\tag{2.134}
$$

$$
\mathbf{\Psi}_r^{\prime\prime(x)} = \begin{bmatrix}
0 & 0 & 0 & 0 & 0 & 0 \\
\Phi_1^{\prime\prime(x)} & \Phi_2^{\prime\prime(x)} & \Phi_3^{\prime\prime(x)} & \Phi_4^{\prime\prime(x)} & \Phi_5^{\prime\prime(x)} & \Phi^{\prime\prime st(x)} \\
0 & 0 & 0 & 0 & 0 & 0
\end{bmatrix},
\tag{2.135}
$$

$$\Psi_r''^{(z)} = \begin{bmatrix} 0 & 0 & 0 & 0 & 0 & 0 \\ \Phi_1''^{(z)} & \Phi_2''^{(z)} & \Phi_3''^{(z)} & \Phi_4''^{(z)} & \Phi_5''^{(z)} & \Phi''^{st(z)} \\ 0 & 0 & 0 & 0 & 0 & 0 \end{bmatrix}, \tag{2.136}$$

where

$$\Psi_r''^{(x)} = \frac{\partial^2 \Psi_r}{\partial x^2}, \qquad \Psi_r''^{(z)} = \frac{\partial^2 \Psi_r}{\partial z^2}. \tag{2.137}$$

The potential field within the piezoelectric plate follows a similar formulation as the one shown in equation (2.122).

As for the beam application, also in this case a matricial formulation can be adopted in order to simplify the practical implementation of this model on a computer. In particular, defining the matrixes $\mathbf{RD}$, $\mathbf{D}_R$ and $\mathbf{C}_R$ as:

$$\mathbf{RD} = \begin{bmatrix} 0 & 1 & 0 & 0 & 0 & 0 \\ 1 & 0 & 0 & 0 & 0 & 0 \\ 0 & 0 & 1 & 0 & 0 & 0 \\ 0 & 0 & 0 & 1 & 0 & 0 \\ 0 & 0 & 0 & 0 & 1 & 0 \\ 0 & 0 & 0 & 0 & 0 & 1 \end{bmatrix}, \tag{2.138}$$

$$\mathbf{D}_R = \begin{bmatrix} \varphi_1''^{(x)} & \varphi_2''^{(x)} & \varphi_3''^{(x)} & \varphi_4''^{(x)} & \varphi_5''^{(x)} & \varphi''^{st(x)} \\ \varphi_1''^{(z)} & \varphi_2''^{(z)} & \varphi_3''^{(z)} & \varphi_4''^{(z)} & \varphi_5''^{(z)} & \varphi''^{st(z)} \\ 0 & 0 & 0 & 0 & 0 & 0 \\ 0 & 0 & 0 & 0 & 0 & 0 \\ 0 & 0 & 0 & 0 & 0 & 0 \\ 0 & 0 & 0 & 0 & 0 & 0 \end{bmatrix}, \tag{2.139}$$

$$\mathbf{C}_R = \upsilon\mathbf{RD} - \mathbf{I}, \tag{2.140}$$

one can write the following equation:

$$N_r(x,z) = \mathbf{L}_u \mathbf{\Psi}_r(x,z) = y\mathbf{C}_R\mathbf{D}_R. \tag{2.141}$$

Finally, by substituting the obtained expressions in equations (2.103)-(2.107), one can get:

$$\mathbf{M}_s = 2h_b\rho_s \int_0^{L_2}\int_0^{L_1} \mathbf{\Psi}_r^T(x,z)\mathbf{\Psi}_r(x,z)\mathrm{d}x\mathrm{d}z, \tag{2.142}$$

$$\mathbf{M}_p = h_p\rho_p \int_{b_1}^{b_2}\int_{a_1}^{a_2} \mathbf{\Psi}_r^T(x,z)\mathbf{\Psi}_r(x,z)\mathrm{d}x\mathrm{d}z, \tag{2.143}$$

$$\mathbf{K}_s = 2\frac{h_b^3}{3} \int_0^{L_2}\int_0^{L_1} \mathbf{D}_R^T(x,z)\mathbf{C}_R^T\mathbf{c}_S\mathbf{C}_R\mathbf{D}_R(x,z)\mathrm{d}x\mathrm{d}z, \tag{2.144}$$

$$\mathbf{K}_p = \left[\frac{(h_b+h_p)^3}{3}-\frac{h_b^3}{3}\right]\int_{b_1}^{b_2}\int_{a_1}^{a_2} \mathbf{D}_R^T(x,z)\mathbf{C}_R^T\mathbf{R}_S^T\mathbf{c}^E\mathbf{R}_S\mathbf{C}_R\mathbf{D}_R(x,z)\mathrm{d}x\mathrm{d}z, \tag{2.145}$$

$$\mathbf{C}_p = \mathbf{N}_v^T(x,z)\mathbf{R}_E^T\mathbf{\varepsilon}^S\mathbf{R}_E\mathbf{N}_v(x,z)\left[h_p(a_2-a_1)(b_2-b_1)\right], \tag{2.146}$$

$$\mathbf{\Theta} = \left[\frac{(h_b+h_p)^2}{2}-\frac{h_b^2}{2}\right]\int_{b_1}^{b_2}\int_{a_1}^{a_2} \mathbf{D}_R^T(x,z)\mathbf{C}_R^T\mathbf{e}^T\mathbf{R}_E\mathbf{N}_V(x,z)\mathrm{d}x\mathrm{d}z, \tag{2.147}$$

where $\rho_s$ and $\rho_p$ are the plate and piezoelectric material densities. The procedure to get to the state-space model is then obtained straight forward as shown in the previous section.

Examples 1 and 2 show how the mathematical model presented in this section can be developed in the particular case of a structure for simulation and analysis purposes. In particular, once that the necessary matrixes are calculated, the implementation of such a model on a computer goes straight forward.

### 2.3.3. Transfer functions for a PVS

The modeling approach shown in the previous section allows to obtain the necessary transfer functions useful at different aims, from identification to control purposes, according to the selected output variable, like the piezoelectric deformation, its electrical charge or the transversal displacement in one point of the structure, and according to the excitations, like a force acting on a point of the structure or the voltage applied to the piezoelectric actuator.

Let us then consider the case of a PVS with a bonded piezoelectric actuator which is driven with a voltage $v$ while an external force $f$ causes vibrations and let us focus on the transversal displacement $y$ of a point of the structure $\mathbf{x}_0$. In order to retrieve the necessary transfer functions, we consider the state-space formulation of equation (2.111). In particular, the matrixes $\mathbf{C}$ and $\mathbf{D}$ are defined as:

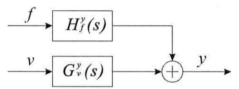

*Figure 2.20 – A schematic diagram for a PVS with voltage v and force f as inputs and transversal displacement y as output.*

$$\mathbf{C}^y = \mathbf{C} = \begin{bmatrix} \mathbf{\Psi}_r(\mathbf{x}_0) & 0 \end{bmatrix}, \qquad \mathbf{D}^y = \mathbf{D} = \begin{bmatrix} 0 & 0 \end{bmatrix}, \qquad (2.148)$$

where the apex $y$ indicates the wanted output variable. Let us then consider the matrix $\mathbf{B}$ partitioned in columns as follows:

$$\mathbf{B} = \begin{bmatrix} \mathbf{B}_{rf} & \mathbf{B}_{rv} \end{bmatrix}, \text{ where } \mathbf{B}_{rf} = \begin{bmatrix} 0 \\ \mathbf{M}^{-1}\mathbf{B}_f \end{bmatrix} \text{ and } \mathbf{B}_{rv} = \begin{bmatrix} 0 \\ \mathbf{M}^{-1}\mathbf{\Theta} \end{bmatrix}. \qquad (2.149)$$

Finally we can write the two transfer functions between the observed transversal displacement and the inputs voltage and force, which are shown in the schematic diagram of Figure 2.20, as:

$$G_v^y(s) = \mathbf{C}^y (s\mathbf{I} - \mathbf{A})^{-1} \mathbf{B}_{rv} + D_v, \tag{2.150}$$

$$H_f^y(s) = \mathbf{C}^y (s\mathbf{I} - \mathbf{A})^{-1} \mathbf{B}_{rf} + D_f, \tag{2.151}$$

where the transfer function pedix indicates the input variable while the apex indicates the output variable, and the matrix $\mathbf{D}^y$ is partitioned as:

$$\mathbf{D}^y = \begin{bmatrix} D_f & D_v \end{bmatrix}, \text{ where } D_f = 0 \text{ and } D_v = 0. \tag{2.152}$$

Let us focus on $G_v^y(s)$, since similar considerations can be made also for $H_f^y(s)$. Such transfer function can also be written as:

$$G_v^y(s) = \frac{\mathbf{C}^y (s\mathbf{I} - \mathbf{A})^* \mathbf{B}_{rv}}{|s\mathbf{I} - \mathbf{A}|} = \frac{n(s)}{d(s)}, \tag{2.153}$$

where $(\ldots)^*$ indicates the adjoint matrix and the order of the two polynomes $n(s)$ and $d(s)$ is depending on the chosen number of modes $N$ to be modeled, even though $d(s)$ has a higher order than $n(s)$ since in this case $D_v$ is zero. Another equivalent representation for such transfer function is given by [RHF03], where this transfer function is expressed as:

$$G_v^y(s) = \sum_{i=1}^{N} \frac{P_i}{s^2 + 2\delta_i \omega_i s + \omega_i^2}, \tag{2.154}$$

where the terms $P_i$ are constants depending on the PVS properties and modal shapes, and $\delta_i$ and $\omega_i$ are respectively the damping factors and natural frequencies.

It is important to remark that the transfer functions obtained with this modeling approach consider a limited number of modes $N$, consequently committing a so-called truncation error. In fact, the direct truncation of higher order modes results into a shift of the zero positions and hence generates errors in the considered frequency range. A typical way to overcome this problem is to add a correction term $K$ which accounts for the higher modes dynamics at the low frequency range:

$$\bar{G}_v^y(s) = G_v^y(s) + K.$$
(2.155)

One way to determine the correction term $K$ consists in considering the static contributions of the higher order modes as follows:

$$K = \sum_{i=N+1}^{M} \frac{P_i}{\omega_i^2},$$
(2.156)

where $M$ should be theoretically infinite to include the infinite number of modes of a structure. More practically one can choose a relatively high number of modes that is also justified by the fact that $\omega_i \to \infty$ per $i \to \infty$, hence very high order modes do not have a significant influence on the low frequency range. More details can be found in [RC00].

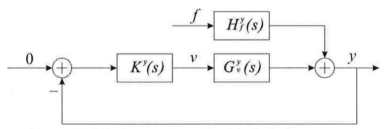

*Figure 2.21 – A typical vibration control loop with position feedback.*

The compensation of the truncation error is very important when the model is needed for tuning a controller for vibration control purposes. Vibration control applications aim to reduce the transversal displacement $y$ by using a controller $K^y(s)$ in a closed control loop configuration, as shown in Figure 2.21. In case the controller is tuned on a model which is not correct unexpectable results can be obtained, which are due to the kind of controller used and to the particular structure.

# 3. Self sensing techniques for piezoelectric materials

In this chapter, self sensing techniques for piezoelectric materials are presented. Such techniques allow to reconstruct the piezoelectric mechanical quantities of interest, i.e. strain and stress, by measuring the electrical ones, i.e. voltage and charge. Thus a piezoelectric transducer can be used simultaneously as an actuator and as a sensor, and for this reason it is called self sensing actuator. Section 3.1 introduces the basic concepts of self sensing. The one step reconstruction and the two steps reconstruction are presented, which allow to reconstruct one or both the mechanical quantities of interest, respectively.

Section 3.2 deals with the dimensioning of self sensing piezoelectric plate actuators, which is a crucial aspect in the design of a self sensing PVS. In fact, it is necessary to find a compromise in the actuator dimensions so that both actuating and sensing capabilities are guaranteed.

Section 3.3 introduces the bridge-circuit based reconstruction, which is the most adopted self sensing technique in scientific literature and it is based on a linear model of the piezoelectric transducer. After a quick survey on this technique its limitations are discussed, and an adaptive algorithm is presented which aims at reducing them.

In section 3.4 a hysteretic dynamic model of a PVS is presented, and the main transfer functions of a PVS with self sensing actuators are defined. Moreover, the effects of the identification error of the piezoelectric electrical capacitance model on the quality of the reconstructed quantities are discussed.

In section 3.5 a hysteretic model based reconstruction algorihm is presented, which makes use of a hysteretic model to perform the reconstruction of the mechanical quantities. In particular, such method is able to perform a self sensing reconstruction both at low and high driving voltages for vibration control purposes.

Finally, section 3.6 deals with the practical implementation of the before mentioned self sensing techniques, which is an important issue since it can affect the performance of a vibration control loop based on self sensing actuators.

## 3.1. Strain and stress reconstruction

Self sensing techniques allow to reconstruct the mechanical quantities of a piezoelectric material as strain $S$ and stress $\sigma$ by measuring the electrical ones, i.e. electric displacement $D$ and electric field strength $E$. Such techniques are based on the knowledge of the piezoelectric behaviour, which is described by its constitutive linear equations, which can be written in four different formulations according to the chosen independent variables.

The constitutive equations (2.2) and (2.4) shown in the previous chapter are written in matricial form, since they describe the piezoelectric material behaviour along all axes. Nevertheless, in order to simplify the discussion, we can scale it down to scalar equations since usually one can individuate a main mechanical axis of interest and the electric field strength of excitation is along the $3^{rd}$ axis. In particular piezoelectric plates (laminar translators) are often used in vibration control applications. Consequently self sensing techniques aim to reconstruct the strain and the stress along the $1^{st}$ or $2^{nd}$ direction. In other applications, instead, where stack translators are utilized (as in the case of the Pendulum Actuator presented in section 2.1.4), the strain and stress of interest is along the $3^{rd}$ axis. In any case, for the following discussion, scalar equations will be used without a particular reference to the axis in order to keep the discussion general with respect to the particular piezoelectric transducer.

### 3.1.1. One step reconstruction

A one step reconstruction aims to reconstruct only one mechanical quantity, either strain or stress. At this aim, only one equation is sufficient to solve the reconstruction.

Let us then recall the sensor equations of equations (2.2) and (2.4) shown in the previous chapter

$$D = d\sigma + \varepsilon^\sigma E, \tag{3.1}$$

$$D = eS + \varepsilon^S E. \tag{3.2}$$

These two sensor equations differ in the choice of the independent variables, that in the first case is the stress $\sigma$ while in the second one is the strain $S$, and in the permittivity, that in the first equation is the *unconstrained permittivity* $\varepsilon^\sigma$ (measured at constant stress) and in the second one is the *clamped permittivity* $\varepsilon^S$ (measured at constant strain). The sensor equation represents the core of self sensing techniques. In fact, by measuring the electric displacement and the electric field strength, one can reconstruct respectively the relative mechanical quantity according to which permittivity (unconstrained or clamped) is used as follows:

$$S_R = \left(D - \varepsilon^S E\right)/e, \tag{3.3}$$

$$\sigma_R = \left(D - \varepsilon^\sigma E\right)/d. \tag{3.4}$$

Both these two mechanical quantities can be then reconstructed by measuring the electrical quantities and by knowing the piezoelectric characteristics, in particular the permittivity (unconstrained or clamped) and the piezoelectric constants, $d$ or $e$. It is of interest that one can reconstruct either the stress or the strain according to the kind of permittivity. Equation (2.6) shows the relation between these two permittivities, which involves also the other piezoelectric material parameters.

Obviously, as discussed above, at the aim of performing a precise reconstruction, the piezoelectric constant as well as the electrical capacitance must be known with high precision. Nevertheless, such parameters influence the reconstruction in a different way. In fact, while an identification error affecting the piezoelectric constants ($d$ or $e$) would result in a gain entailing a shift of the amplitude diagram, an error affecting the electrical capacitance can result in a wrong phase and amplitude reconstruction, which can lead vibration control loops based on the reconstructed signal as feedback to instability.

### 3.1.2. Two steps reconstruction

A two steps reconstruction aims to reconstruct both the mechanical quantities of interest, i.e. strain as well as stress, by using both the sensor and actuator equations.

For simplicity, let us consider only the constitutive equations of a piezoelectric material expressed in equation (2.2) in the scalar form:

$$D = d\sigma + \varepsilon^\sigma E$$
$$S = s^E \sigma + dE. \tag{3.5}$$

Starting from these equations, the following reconstructor can be defined:

$$\sigma_R = \left(D - \varepsilon^\sigma E\right)\big/d$$
$$S_R = s^E \sigma_R + dE, \tag{3.6}$$

where the first reconstruction step consists in using the sensor equation to obtain the reconstructed stress $\sigma_R$ which is then used for calculating the reconstructed strain $S_R$ in the second reconstruction step. Similarly one can invert the order of reconstruction by considering the constitutive equations shown in (2.4), where the strain is reconstructed at the first step and the stress at the second one.

As one can see such reconstruction is strongly based on the knowledge of the piezoelectric parameters, and a reconstruction error obtained at the first step is then propagated to the second one.

Let us consider the case of the piezoelectric plate of dimensions *Lxbxh* shown in Figure 3.1, where the mechanical quantities of interest, stress and strain, are along the 1$^{st}$ axis and the polarization along the 3$^{rd}$ one.

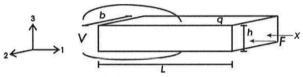

*Figure 3.1 – Piezoelectric plate and related quantities.*

The small signal constitutive equations can now be expressed in terms of force *F*, voltage *V*, charge *q* and elongation *x* (time dependence is here neglected):

$$x = \frac{Ld_{31}}{h}V + \frac{L}{A_f}s_{11}^E F$$

$$q = \frac{A_e \varepsilon_{33}^\sigma \varepsilon_0}{h}V + \frac{A_e d_{31}}{A_f}F, \qquad (3.7)$$

where $A_e$ is the area of the electrode ($A_e = Lb$) and $A_f$ is the area subjected to the force $F$ ($A_f = bh$). With the transfer to larger amplitudes the nonlinear processes are excited. This entails a hysteretic transfer behaviour which is described with the help of hysteretic operators [KP89]. It is then possible to rewrite the equation (3.7) as follows [JK06]:

$$x = \Gamma_A[V] + \Gamma_M[F]$$

$$q = \Gamma_E[V] + \Gamma_S[F]. \qquad (3.8)$$

Here $\Gamma_A$, $\Gamma_M$, $\Gamma_E$ and $\Gamma_S$ are general hysteretic operators. The mechanical quantities are then reconstructed according to the equations

$$F_R = \Gamma_S^{-1} \left[ q - \Gamma_E \left[ V \right] \right]$$
$$x_R = \Gamma_A \left[ V \right] + \Gamma_M \left[ F_R \right].$$

(3.9)

Let us consider a real piezoelectric plate, in particular a PIC 151 produced by PI Ceramic, whose parameters of interest are shown in Table 3.1.

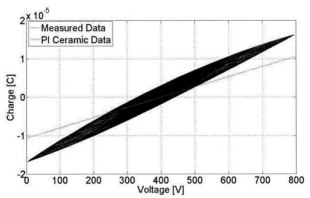

*Figure 3.2 – Measured large signal characteristic $\Gamma_E$ compared with the linear characteristic from PI Ceramic.*

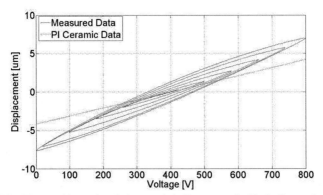

*Figure 3.3 – Measured large signal characteristic $\Gamma_A$ compared with the linear characteristic from PI Ceramic.*

Equipped with a voltage amplifier capable of driving a capacitive load in the range of $0...800$ V, the measured characteristics $\Gamma_E \left[ V \right]$ and $\Gamma_A \left[ V \right]$ are shown in the Figures 3.2 and 3.3 and compared with the linear characteristics based on the data provided from the producer of the piezoelectric plate.

In Figures 3.2 and 3.3 one can notice that the linear characteristics have a very similar slope with respect to the measured ones only within a small driving signal range.

In order to measure the remaining two characteristics $\Gamma_S$ and $\Gamma_M$, a guiding support has been realized (Figure 3.4a) and inserted into a measuring set up (Figure 3.4b).

The measured characteristics $\Gamma_S[F]$ and $\Gamma_M[F]$ are shown in Figures 3.5 and 3.6. As one can see the characteristics $\Gamma_S[F]$ and $\Gamma_M[F]$ have turned out to be linear in the chosen range, while $\Gamma_E[V]$ and $\Gamma_A[V]$ are strongly hysteretic.

Let us assume that the available models for these last two characteristics are affected respectively by the errors $e_{qv}$ and $e_{xv}$, where one can define the modeling error $e$ of a general hysteretic operator $\Gamma[\cdot]$ as:

$$e = \frac{\|\Gamma[\cdot](t) - y(t)\|_\infty}{\|\Gamma[\cdot](t)\|_\infty}, \tag{3.10}$$

where $y(t)$ is the output measurement data.

| Parameter Symbol | Value |
|---|---|
| $\varepsilon_{33}^\sigma / \varepsilon_0$ | 2400 |
| $d_{31}$ | $-210 \cdot 10^{-12}$ C / N |
| $s_{11}^E$ | $15 \cdot 10^{-12}$ m$^2$ / N |
| $L$ | 50 mm |
| $b$ | 30 mm |
| $h$ | 1 mm |

*Table 3.1 – Parameters of the piezoelectric plate PIC151.*

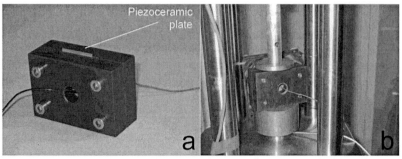

*Figure 3.4 – Testing set up for piezoelectric plate characterization. a) A guiding support for the piezoelectric plate, b) measuring set up for force application and displacement measurements.*

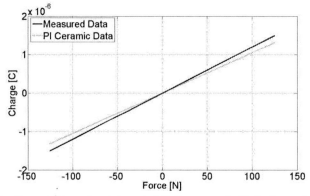

*Figure 3.5 – Measured large signal characteristic $\Gamma_S$ compared with the linear characteristic from PI Ceramic.*

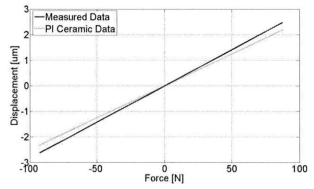

*Figure 3.6 – Measured large signal characteristic $\Gamma_M$ compared with the linear characteristic from PI Ceramic.*

Considering that equation (3.8) can be rewritten as [GPJ10]:

$$q = q_V + q_F$$
$$x = x_V + x_F,$$

(3.11)

where the contributions of voltage and force on the charge and on the elongation are highlighted. Due to the modeling errors of the hysteretic characteristics, the amount of charge $q_V$ obtained from the driving voltage as well as the displacement in unloaded condition $x_V$ can be expressed as:

$$\hat{q}_V = q_V \pm \tilde{q}_V$$
$$\hat{x}_V = x_V \pm \tilde{x}_V,$$

(3.12)

where the apex $\sim$ indicates the measurement uncertainty and the apex $\wedge$ indicates the measured quantity.

Consequently, the reconstructed quantities $\hat{F}_R$ and $\hat{x}_R$ are also affected by error as follows:

$$\hat{F}_R = F \pm \tilde{F}_R$$
$$\hat{x}_R = x \pm \tilde{x}_R.$$

(3.13)

Since the characteristics $\Gamma_S[F]$ and $\Gamma_M[F]$ can be considered with good approximation to be linear, it makes sense to use the linear equations to obtain:

$$\tilde{F}_R = e_{qv} A_f \frac{\varepsilon_{33}^\sigma \varepsilon_0}{d_{31}} \frac{V}{h}$$
$$\tilde{x}_R = L \left( e_{qv} \frac{s_{11}^E \varepsilon_{33}^\sigma \varepsilon_0}{d_{31}} + e_{xv} d_{31} \right) \frac{V}{h}.$$

(3.14)

The amount of modeling error is propagated up to the reconstructed strain. In the case of a linear reconstruction, the two modeling errors considered become larger at increasing the driving voltage, while the hysteretic models keep this error constant (between 1% and 2%) almost independently on the magnitude of the driving voltage.

Moreover, equation (3.14) promotes the choice of the geometrical and physical parameters of the piezoelectric actuator for a particular self-sensing application, in order to reduce the uncertainty on the reconstructed quantities.

## 3.2. Dimensioning of self sensing piezoelectric plates

Self sensing actuators aim to replace the typical configuration of collocated actuators and sensors, where actuators and sensors share the same deformation of the structure. Consequently it makes sense to investigate deeper into such configuration.

In particular, the PVS models derived in section 2.3.2 have been developed based on the Euler-Bernoulli beam theory. In many cases involving collocated or nearly collocated actuators and sensors such theory is inappropriate, since the transfer function of a collocated sensor/actuator becomes particularly sensitive to the strain transmission path within the coupled system, and the location of the open-loop transmission zeros can vary significantly [Pre02]. Consequently a more detailed theory must be used that can account for membrane (or in-plane) deformations as well as bending ones. In fact, in conditions where the structure is particularly thin with respect to the piezoelectric plates thicknesses while having similar Young's modulus, the membrane strains play an important role in the transmission of the strain from the actuator to the sensor.

Considering the case of a beam, one can notice from Figure 2.17 that a piezoelectric plate actuator has to have a thickness around 1/2 or 1/3 times the thickness of the structure at same Young's modulus in order to induce the highest possible curvature and consequently achieve a good actuation authority. Nevertheless, as one will see further in this text, such condition is not always desirable for self sensing applications, since it leads to small local strains and consequently to a low amount of mechanical charge for an accurate self sensing reconstruction.

The limitations of the classical beam theory can be handled by the shell theory [Lee90]. In fact, according to the shell (plate) theory [Jon75], [And86], in the case of a laminated plate, the total strain can be expressed as:

$$\mathbf{S} = \mathbf{S}_0 + y\mathbf{\kappa}, \tag{3.15}$$

where $\mathbf{S}_0$ is the deformation vector of the mid-plane and $\mathbf{\kappa}$ is the vector of curvatures. In particular, $\mathbf{S}_0$ is dependent on the actuation strain $\Lambda$ shown in equation (2.67), which is proportional to the voltage applied to the piezoelectric actuator. This is then going to lead to the introduction of transmission zeros in the final transfer function. For the case of an isotropic plate and one piezoelectric actuator, the vector $\mathbf{S}_0$ can be defined as:

$$\mathbf{S}_0 = \begin{bmatrix} k_t \\ k_t \\ 0 \\ 0 \\ 0 \\ 0 \end{bmatrix} \Lambda = \mathbf{K}_v V, \qquad k_t = \frac{2}{2 + \Psi_e}, \tag{3.16}$$

where $\Psi_e$ is defined in equation (2.75) and $V$ is the voltage applied to the piezoelectric actuator.

Under these assumptions, the dynamic model proposed in section 2.3.2 can be modified by expressing the strain vector $\mathbf{S}$ as:

$$\mathbf{S} = \mathbf{K}_v(\mathbf{x})\mathbf{v}(t) + N_r(\mathbf{x})\mathbf{r}(t), \tag{3.17}$$

where the matrix $\mathbf{K}_v(\mathbf{x})$ depends on the spatial coordinates, since it is equal to zero where no piezoelectric materials are bonded and dependent on the particular piezoelectric actuator where these are instead bonded.

After some calculation, one can obtain the following actuator and sensor equations for a PVS:

$$\left(\mathbf{M}_s + \mathbf{M}_p\right)\ddot{\mathbf{r}} + \left(\mathbf{K}_s + \mathbf{K}_p\right)\mathbf{r} - \mathbf{\Theta}_A \mathbf{v} = \mathbf{B}_f \mathbf{f}$$
$$\mathbf{\Theta}_A^T \mathbf{r} + \mathbf{C}_p \mathbf{v} + \mathbf{\Theta}_B \mathbf{v} = \mathbf{B}_q \mathbf{q},$$

$$(3.18)$$

where

$$\mathbf{\Theta}_A = \mathbf{\Theta} - \mathbf{K}_{pv}, \qquad (3.19)$$

$$\mathbf{K}_{pv} = \int_{V_p} N_r^T(\mathbf{x}) \mathbf{R}_r^T \mathbf{c}^E \mathbf{R}_s \mathbf{K}_v(\mathbf{x}), \qquad (3.20)$$

$$\mathbf{\Theta}_B = 2\int_{V_p} \mathbf{K}_v^T(\mathbf{x}) \mathbf{R}_s^T \mathbf{e}^T \mathbf{R}_E N_v(\mathbf{x}) - \int_{V_p} \mathbf{K}_v^T(\mathbf{x}) \mathbf{R}_s^T \mathbf{c}^E \mathbf{R}_s \mathbf{K}_v(\mathbf{x}). \qquad (3.21)$$

The quantities used in equations (3.18) - (3.21) have been defined in section 2.3.2.

In particular, the term $\mathbf{\Theta}_B \mathbf{v}$ introduces transmission zeros in the transfer functions obtained from the above equations. This has the effect of taking the zeros closer to the poles, and thus they reduce the frequency response dynamics.

Generally piezoelectric actuators are chosen for their actuating properties, and thus they need to have a considerable thickness (as discussed in section 2.3.1) and Young's modulus. On the other side, piezoelectric sensors need to be very thin not to influence significantly the structural deformation and obtain a high resolution sensory information. Self sensing actuators are required to perform both actions at the same time, and for this reason a compromise has to be found between actuating and sensing capabilities. Let us recall the parameter $\Psi_e$ defined in equation (2.66):

$$\Psi_e = \frac{E_s A_s}{E_a A_a}.$$

(3.22)

Choosing piezoelectric transducers which keep this parameter equal to 10 is by experience a good choice. In fact, in this way, the self sensing piezoelectric actuator will still preserve good actuating capabilities without sacrificing excessively its role as a sensor.

## 3.3. Bridge-circuit based reconstruction

In the field of vibration control, one of the first methods for implementing a self sensing piezoelectric plate was proposed in [AH94]. Such method realizes a one step reconstruction of the strain by using an electrical bridge circuit, whose principal schematic is shown in Figure 3.7.

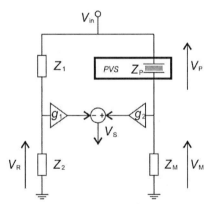

*Figure 3.7 – A self sensing impedance bridge circuit for strain measurement.*

As one can see, the piezoelectric actuator bonded on the structure, which is electrically represented by the impedance $Z_P$, is inserted in a brige circuit, whose output voltage $V_S$ is proportional to the mechanically induced charge when the bridge is correctly tuned. In fact, the voltage $V_{in}$ driven by the voltage amplifier is then divided in the right arm of the bridge in order to leave a useful piezoelectric driving voltage $V_P$ which amounts to:

$$V_P = \frac{Z_P}{Z_P + Z_M}\left(V_{in} - sZ_2\mathbf{\Theta}^T\mathbf{r}\right) \approx \frac{Z_P}{Z_P + Z_M}V_{in} \triangleq \beta V_{in},$$

(3.23)

where the approximation is valid if the input voltage $V_{in}$ is much larger than the one measured by the sensing effect, which is typical for such kind of applications.

Let us then consider also that the impedances $Z_1$ and $Z_2$ have equal charger as well as the impedances $Z_P$ and $Z_M$. In order to get the sensor voltage, the voltages across the impedances $Z_2$ and $Z_M$ are multiplied by the gain $g_1$ and $g_2$ and subtracted. The voltage $V_R$ across the impedance $Z_2$ and the voltage $V_M$ across the impedance $Z_M$ can be expressed as:

$$V_R = g_1 \frac{Z_2}{Z_2 + Z_1}V_{in},$$

(3.24)

$$V_M = g_2 \frac{Z_M Z_P}{Z_M + Z_P}s\left(\frac{V_{in}}{sZ_P} + \mathbf{\Theta}^T\mathbf{r}\right).$$

(3.25)

The sensor voltage $V_S$ is then obtained by subtracting these two voltages:

$$V_S = \left(g_1 \frac{Z_2}{Z_2 + Z_1} - g_2 \frac{Z_M}{Z_M + Z_P}\right)V_{in} - sg_2 \frac{Z_M Z_P}{Z_M + Z_P}\mathbf{\Theta}^T\mathbf{r} \triangleq$$
$$Z_v(s)V_{in} - Z_m(s)\mathbf{\Theta}^T\mathbf{r}.$$

(3.26)

When the bridge is balanced, $Z_v(s) = 0$ and the sensor voltage $V_S$ is proportional only to the mechanical strain $\mathbf{\Theta}^T\mathbf{r}$. Such condition can be achieved by keeping a constant gain $g_2$ and varying the gain $g_1$ or by varying the impedance $Z_M$. Such impedance practically consists of a capacitor with a parallel resistor.

## 3.3.1. Limits of a typical bridge based self sensing technique

The kind of circuit discussed above presents some issues that can compromise its performance that concerns mostly three aspects:

- Tuning of the components or bridge balancing
- Dynamic behaviour
- Non modeled nonlinear effects

As observed before, the bridge must be perfectly balanced in order to provide a reliable measurement. Nevertheless, considering that the impedances are realized as a capacitor in parallel with a resistor, one has to know with high precision the values of the clamped capacitance and of the leakage resistance of the piezoelectric actuator. In fact, even a small error affecting the balancing of the bridge can lead to a high measurement error, since the mechanical charge to be extracted is usually some few percents of the whole charge. In particular the value of the clamped capacitance affects the balancing of the bridge at high frequencies, while the value of the leakage resistance has more effect at low frequencies and it is related to the dynamic behaviour issue.

The value of the clamped capacitance is usually calculated by using the formula of equation (2.6) and by considering the geometry of the piezoelectric actuator. Indeed, a direct measurement of this parameter is highly difficult, since it requires to perform a measurement while constraining the piezoelectric actuator to zero strain. Consequently this parameter is typically affected by identification errors. A fine tuning of the gain $g_1$ is then required which is performed with a trial and error approach.

Another issue to be faced is the dynamic behaviour of the bridge. In fact, both branches of the circuit are made of resistances/capacitances in parallel, and consequently each one has its own dynamics and it is important that the dynamis of both branches have the same, or at least very similar, behaviour, especially in terms of phase. By using the notation $Z_X = (C_X \| R_X)$, the dynamic balancing is reached when the following relation is satisfied:

$$\frac{R_P C_P}{R_M C_M} = \frac{R_1 C_1}{R_2 C_2}. \tag{3.27}$$

The dynamic balancing of the bridge concerns the right knowledge of the leakage resistance value, which can only be vaguely estimated. It is then opportune to tune the bridge in a way that the operative frequency range is far from the low cut-off frequency of the bridge.

Finally, the most important aspect of the bridge circuit concerns the nonlinearity of the piezoelectric actuators. In fact, the electrical clamped capacitance is affected by rate independent hysteresis as well as creep effects. Both nonlinearities concur to a systematic error in the measurement signal due to the fact that the term $Z_v(s)$ of equation (3.26) is not zero. Indeed, the reference capacitance has a linear behaviour that can not account for such nonlinearities. For such reason, the bridge circuit has good performance only at low driving voltages, where the hysteretic loops of the electrical clamped capacitance are particularly narrow and so quasi linear. The higher the driving voltage the higher the measurement error. This issue is common to all methods based on linear modeling.

To summarize, such approach has the advantage of being easy to be implemented electrically. Nevertheless it performs well only at very low driving voltages and the tuning procedure can be burdensome.

### 3.3.2. Adaptive algorithms for self sensing techniques

The identification of the piezoelectric clamped capacitance is a critical issue since it is crucial for the strain reconstruction and also very difficult to identify with a direct measurement. Indeed, small errors in the capacitance identification can lead to high reconstruction errors. At this purpose, an interesting contribution comes from the work of Cole and Clark, [CC94], where an adaptive algorithm is proposed.

Let us consider a PVS with $n$ piezoelectric actuators bonded on it and $m$ structural modes. The electrical charge stored in the $k^{th}$ actuator for the multi-input/single-output case, which is related to the applied voltage $v_k$ and the mechanical excitation of the structure, can be rewritten as:

$$q_k = C_{p,k}v_k + \sum_{i=1}^{n}\sum_{j=1}^{m}\frac{\Theta_{ij}\Theta_{kj}}{s^2 + 2\zeta_j\omega_j s + \omega_j^2}v_i. \tag{3.28}$$

Converting the transfer function expressed in equation (3.28) from the Laplace domain to the $z$-domain, and assuming a zero-order hold, gives:

$$q_k(z) = C_{p,k}v_k(z) + \sum_{i=1}^{n}\sum_{j=1}^{m}\frac{b_{j1}^{ki}z + b_{j0}^{ki}}{z^2 + a_{j1}^{ki}z + a_{j0}^{ki}}v_i(z) =$$

$$C_{p,k}v_k(z) + \sum_{i=1}^{n}H_{ki}(z)v_i(z), \tag{3.29}$$

where the coefficients $b_0$, $b_1$, $a_0$ and $a_1$ are related to the continuous time parameters and the sampling period. Converting now equation (3.29) to the time domain gives:

$$q_k[t] = C_{p,k}v_k[t] + \sum_{i=1}^{n}\sum_{j=-\infty}^{+\infty}h_{ki}[j]v_i[t-j], \tag{3.30}$$

where $h_{ki}$ is the structural response in function of the time that relates the mechanical charge stored in the $k^{th}$ piezoelectric actuator to the impulsive excitation of the $i^{th}$ piezoelectric actuator. In particular, the first term of equation (3.30) is the electrical response provided by the piezoelectric capacitance, while the second term is the mechanical response.

An adaptive filter is a discrete filter whose weights are changed by an adaptive algorithm, and they are successfully used in identification problems [Hay91]. In general, an adaptive filter of order $M$ is the weighted sum of the last $M$ samples of the input. A first order adaptive filter can be used to identify the electrical clamped capacitance: the output of the filter, in fact, will depend only on the current value of the input and not on older samples.

Let us then consider the total charge to be the desired signal of the filter, the driving voltage to be the input of the filter and the piezoelectric capacitance its only weight. Consequently, the error signal which drives the adaptive algorithm is the mechanical charge $q_m$ as shown in Figure 3.8.

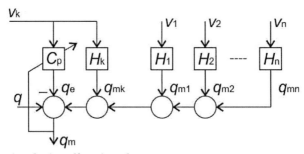

*Figure 3.8 – An adaptive self sensing scheme.*

In particular, the LMS (Least Mean Squares) and RLS (Recursive Least Squares) adaptive filters are here discussed.

Using the LMS algorithm, the adaptive filter weight converges in the mean to the optimal Wiener filter solution, which satisfies:

$$\mathbf{w}_{opt} = \mathbf{R}^{-1}\mathbf{p},$$

(3.31)

where $\mathbf{R}$ is the input correlation matrix for the filter input $v_k$ and $\mathbf{p}$ is the cross correlation between the filter input and desired signal $q_k$. These two quantities can be expressed as:

$$\mathbf{R} = E\{v_k[t]v_k[t]\} = \sigma_k^2,$$

(3.32)

$$\mathbf{p} = E\{v_k[t]q_k[t]\} = C_{p,k}\sigma_k^2 + E\{v_k[t]q_{m,k}[t]\}. \tag{3.33}$$

Consequently the optimal Wiener filter for the $k$ piezoelectric actuator is:

$$w_{opt,k} = C_{p,k} + \frac{E\{v_k[t]q_{m,k}[t]\}}{\sigma_k^2}. \tag{3.34}$$

In case the mechanically induced charge of the $k^{\text{th}}$ piezoelectric actuator is not correlated to its driving voltage, the second term of equation (3.34) is zero. Thus the Wiener filter weight is:

$$w_{opt,k} = C_{p,k}. \tag{3.35}$$

The filter converges to the right value of the capacitance only if the mechanicaly induced charge and the voltage are not correlated, which occurs either if the input voltage is white noise or if the driving voltage is 90° shifted respect to the mechanical charge. If the driving voltage and the mechanical charge are correlated, the identification will contain a bias error that would corrupt the self sensing reconstruction.

Such results are common both to the LMS and to the RLS algorithms. Nevertheless, the RLS algorithm has a quicker convergence time even though computationally more complex. Moreover the adaptive identification ignores parametrical uncertainties, i.e. there is no need to perform an identification of the parameters of the measuring circuit since the identified value of the piezoelectric capacitance will compensate any uncertainty.

### 3.3.3. Limits of the linear piezoelectric capacitance model

All self sensing techniques presented up to this point make use of the linear constitutive equations of a piezoelectric material to perform the reconstruction of the mechanical quantities. The main parameter needed at this aim is the electrical clamped capacitance, and, as shown in section 2.1.3, it is hysteretic although it has a linear behaviour only at very low electrical excitation. It is of interest to observe how a linear self sensing reconstruction performs at different driving voltages. Let us consider a piezoelectric plate with a thickness of 1 mm, which can be driven up to 1 kV at full driving voltage, bonded on a cantilevered beam and let us drive it at a frequency far from any resonances or antiresonances, where the phase shift between the deformation of the piezoelectric plate and the driving voltage is assumed to be zero. One can observe in Figure 3.9 that, as expected, the reconstructed strain of the actuator is rather in phase with the driving voltage when this is in a range of ±12 V.

In fact, the hysteretic loop of the electrical capacitance is so small that it can be considered linear at a very good approximation. The electrical capacitance (red line) used for reconstruction has been identified by using a LMS adaptive algorithm.

In Figure 3.10, instead, a driving voltage in the range ±75 V is applied and the strain is reconstructed by using the same algorithm as the previous example.

One can observe how the hysteresis loop has become bigger, and this leads to an erroneous phase reconstruction, which is in this example around 20°. Such driving voltage is still fairly low respect to the maximum driving voltage, but it is already big enough to introduce significant error in the phase reconstruction. In fact, the hysteretic component of the electrical capacitance can not be compensated by the linear model used by the self sensing reconstruction. Thus, as mentioned previously, such a consistent phase shift would lead vibration controllers to instability.

Since the reconstructed strain is typically fed back to a vibration controller, any errors in the phase reconstruction would contribute to a loss of stability of the controller or to a reduction of the control capabilities and consequently the possibility to damp stronger vibrations. Such limit entails the need of a more accurate reconstruction, which can be obtained exclusively by keeping into account the hysteretic component of the electrical capacitance in order to keep the phase errors within acceptable ranges at all driving voltages.

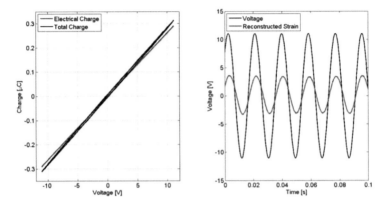

*Figure 3.9 – Reconstructed strain based on permittivity modeling at low driving voltage. a) Capacitance characteristic; b) driving voltage and reconstructed strain vs. time.*

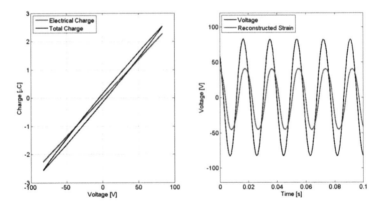

*Figure 3.10 – Reconstructed strain based on permittivity modeling at high driving voltage a) Capacitance characteristic; b) driving voltage and reconstructed strain vs. time.*

## *3.4. Hysteretic dynamic model of a PVS*

The dynamic model of a PVS described in section 2.3.2 (linear dynamic model) considers the piezoelectric transducer within its linear range of operation. Indeed, the linear constitutive equations of a piezoelectric material are used. It is then of interest to extend such model to consider also the piezoelectric nonlinearities and the way they affect the dynamics of a PVS.

The starting point for modeling a PVS are the piezoelectric constitutive equations, that in the case of the linear dynamic model are the ones shown in equation (2.93). Let us define a generic hysteretic operator as $y = [x]$, where $x$ and $y$ are respectively its input and its output. One can now define a set of hysteretic constitutive equations as:

$$\begin{bmatrix} \mathbf{D}' \\ \mathbf{\sigma}' \end{bmatrix} = \begin{bmatrix} \mathbf{\varepsilon}^S[\cdot] & \mathbf{g}[\cdot] \\ -\mathbf{e}[\cdot] & \mathbf{c}^E \end{bmatrix} \begin{bmatrix} \mathbf{E}' \\ \mathbf{S}' \end{bmatrix}, \tag{3.36}$$

where the permittivity matrix $\mathbf{\varepsilon}^S$ and the electro-mechanical matrix $e$ are now expressed as the hysteretic operators $\mathbf{\varepsilon}^S[\cdot]$ and $\mathbf{e}[\cdot]$. In particular, in this case the quantities $e$ and $\mathbf{e}^T$ of equation (2.93) become respectively the hysteretic operators $\mathbf{g}[\cdot]$ and $\mathbf{e}[\cdot]$, since the transposition operation has no meaning in this case. Moreover, the piezoelectric stiffness $\mathbf{c}^E$ can still be considered as a linear characteristic even for broad stress or strain ranges. Such equations give a more reliable description of the real behaviour of a piezoelectric transducer than the linear equations (2.93) since they consider the hysteretic nature of the piezoelectric transducer.

By using equation (3.36) in place of equation (2.93) in the derivation of the dynamic model, the actuator and sensor equations obtained can be written as:

$$\left(\mathbf{M}_s + \mathbf{M}_p\right)\ddot{\mathbf{r}} + \left(\mathbf{K}_s + \mathbf{K}_p\right)\mathbf{r} - \mathbf{\Theta}_v[\mathbf{v}] = \mathbf{B}_f\mathbf{f},$$
$$\mathbf{\Theta}_r[\mathbf{r}] + \mathbf{C}_p[\mathbf{v}] = \mathbf{B}_q\mathbf{q}, \tag{3.37}$$

where the electro-mechanical coupling matrix $\mathbf{\Theta}$ is now a matrix of

90

hysteretic operators.

From the actuator and sensor equations it is possible to obtain the state space representation in the same way of equation (2.111), that is here recalled:

$$\dot{\xi} = A\xi + B\tau$$
$$\chi = C\xi + D\tau.$$
(3.38)

The state matrix $A$ keeps the same representation of the linear case, while the forcing matrix $B$ contains now a hysteretic operator and is expressed as:

$$B = \begin{bmatrix} 0 & 0 \\ -M^{-1}B_f & M^{-1}\Theta_v[\cdot] \end{bmatrix}.$$
(3.39)

As one can see, the system remains linear except for the forcing matrix $B$ which contains the hysteretic operator $\Theta_v[\cdot]$ applying to the voltage inputs.

One can then consider a PVS as a linear system where the forces/moments introduced by the piezoelectric materials are hysteretic. Thus similar considerations can be made both in the case of a linear model and of a hysteretic one.

## 3.4.1. Transfer functions for a PVS with self sensing actuators

Self sensing techniques for vibration control applications aim at extracting the mechanically induced charge on a piezoelectric actuator in order to be fed back to the vibration controller. Recalling the sensor equation of equations (2.102) and (3.37) (respectively related to the linear and to the hysteretic models), one can consider that the total charge $q$ stored on a piezoelectric tranducer of a PVS is the sum of the mechanically induced charge $q_m$ (respectively $q_m = \Theta^T r$ and $q_m = \Theta_r[r]$) and the electrically generated charge $q_e$ (respectively $q_e = C_p v$ and $q_e = C_p[v]$). The mechanical charge quantity is the sensor charge, since it is induced by the deformation of the transducer. Let us then consider the case of a PVS with one piezoelectric actuator which is intended to be used as a self sensing actuator for vibration control. Moreover, let us assume that the mechanical charge $q_m$ is directly measureable and fed back to the controller $K^{q_m}(s)$ as shown in Figure 3.11.

The transfer functions $G_v^{q_m}(s)$ and $H_v^{q_m}(s)$ are defined as:

$$G_v^{q_m}(s) = \mathbf{C}^{q_m}(s\mathbf{I} - \mathbf{A})^{-1}\mathbf{B}_{rv}, \tag{3.40}$$

$$H_f^{q_m}(s) = \mathbf{C}^{q_m}(s\mathbf{I} - \mathbf{A})^{-1}\mathbf{B}_{rf}, \tag{3.41}$$

where the matrix $\mathbf{B}$ has been partitioned as shown in equation (2.149) and the output matrix is defined for the linear case as:

$$\mathbf{C}^{q_m} = \begin{bmatrix} \mathbf{\Theta}^T & \mathbf{0} \end{bmatrix}, \tag{3.42}$$

or, in the case of the hysteretic model, as:

$$\mathbf{C}^{q_m} = \begin{bmatrix} \mathbf{\Theta}_r[\cdot] & \mathbf{0} \end{bmatrix}. \tag{3.43}$$

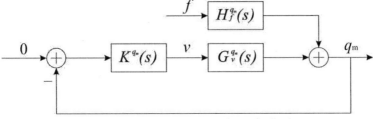

*Figure 3.11 – Vibration control loop with mechanical charge feedback.*

Such control schema is very similar to the one shown in Figure 2.21. In fact, the mechanical charge $q_m$ is proportional to the piezoelectric deformation, which is related to the transversal displacement $y$ by the modal shapes. However, while the transversal displacement is practically measureable with a position sensor, the mechanical charge can not be measured directly. In order to estimate it, a self sensing reconstructor unit is used. Let us define a reconstructor $R$ as follows:

$$\hat{q}_m = R(v,q), \qquad\qquad (3.44)$$

where $\hat{q}_m$ is the reconstructed mechanical charge obtained by the reconstructor. Thus, a typical vibration control loop based on a self sensing actuator is shown in Figure 3.12.

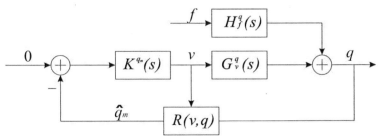

*Figure 3.12 – Vibration control loop with reconstructed mechanical charge feedback.*

In this case, the transfer functions $G_v^q(s)$ and $H_v^q(s)$ are defined as:

$$G_v^q(s) = \mathbf{C}^q(s\mathbf{I} - \mathbf{A})^{-1}\mathbf{B}_{rv} + \mathbf{D}^q, \qquad\qquad (3.45)$$

$$H_v^q(s) = \mathbf{C}^q(s\mathbf{I} - \mathbf{A})^{-1}\mathbf{B}_{rf} + \mathbf{D}^q, \qquad\qquad (3.46)$$

where the output matrixes are defined for the linear case as:

$$\mathbf{C}^q = \begin{bmatrix} \mathbf{\Theta}^T & 0 \end{bmatrix}, \qquad \mathbf{D}^q = \begin{bmatrix} 0 & C_p \end{bmatrix}, \qquad\qquad (3.47)$$

or for the hysteretic case as:

$$\mathbf{C}^q = \begin{bmatrix} \mathbf{\Theta}_r[\cdot] & 0 \end{bmatrix}, \qquad \mathbf{D}^q = \begin{bmatrix} 0 & C_p \end{bmatrix}. \qquad\qquad (3.48)$$

The reconstructed mechanically induced charge $\hat{q}_m$ is an estimation of the mechanical charge, and, thus, can be expressed as:

$$\hat{q}_m = q_m + \tilde{q}_m,\tag{3.49}$$

where $\tilde{q}_m$ is the reconstruction error. Let us consider a linear reconstructor, which is defined as:

$$R(v,q) = q - \hat{C}_p v .\tag{3.50}$$

In this case the capacitance identification error $\tilde{C}_p$ is defined as:

$$\tilde{C}_p = \hat{C}_p - C_p .\tag{3.51}$$

It is now important to remark that the capacitances in equation (3.51) are assumed to be linear. Let us now consider the case that no forces are acting on the PVS, i.e. $f = 0$. By considering equations (3.49) and (3.51), the transfer function between the reconstructed mechanical charge and the voltage can be then defined as:

$$G_v^{\hat{q}_m}(s) = \mathbf{C}^q \left(s\mathbf{I} - \mathbf{A}\right)^{-1} \mathbf{B}_{rv} + \tilde{C}_p ,\tag{3.52}$$

where $\mathbf{C}^q = \mathbf{C}^{q_m}$. This equation can be more extensively rewritten as:

$$G_v^{\hat{q}_m}(s) = \frac{\mathbf{C}^q \left(s\mathbf{I} - \mathbf{A}\right)^* \mathbf{B}_{rv}}{\left|s\mathbf{I} - \mathbf{A}\right|} + \frac{\tilde{C}_p \left|s\mathbf{I} - \mathbf{A}\right|}{\left|s\mathbf{I} - \mathbf{A}\right|} = \\ G_v^{q_m}(s) + \tilde{C}_p .\tag{3.53}$$

The reconstructed transfer function of the PVS is then equal to the real one plus a static contribution which is the capacitance identification error $\tilde{C}_p$. This term has an effect on the position of the zeros. In fact, as one can notice from equation (3.53), the poles of the reconstructed transfer function do not move, while the zeros are shifted according to the entity of the capacitance identification error. In particular, zeros are shifted towards the poles as the modeling error increases, and, in case such error is particularly high, a zeros/poles cancellation can occur. Consequently, the reconstructed transfer function $G_v^{\hat{q}_m}(s)$ tends to a flat amplitude and phase behaviour as the capacity modeling error increases. Obviously, the sensing signal which is fed back to the vibration controller brings erroneous information, with the consequence of limited control performance or even instability.

## 3.5. Hysteretic model based reconstruction

The linear modeling of the electrical clamped capacitance results into significant reconstruction error because of its hysteretic nature. The modeling error in equation (3.51) is not accurate since it considers the real capacitance to be linear. Consequently, a linear reconstructor is rather affected by the following capacitance error:

$$\tilde{C}_p[\cdot] = \hat{C}_p - C_p[\cdot], \tag{3.54}$$

where the real electrical clamped capacitance is now expressed as a hysteretic model. Also the modeling error is now expressed as a hysteretic model of the input voltage, since it is the difference between the real hysteretic behaviour and its linear approximation. By assuming that the measured charge is exclusively the electrical charge, i.e. the measured mechanical charge is totally uncorrelated to the electrical one, a simulation example is shown in Figure 3.13.

As one can see, although the peak to peak amplitude of the real characteristic is well matched by the linear model, the capacitance modeling error $\tilde{C}_p[\cdot]$ is more consistent at the driving voltage crossing zero, which implies a wrong phase shift of the reconstructed signal and consequently instability problems for the vibration controller.

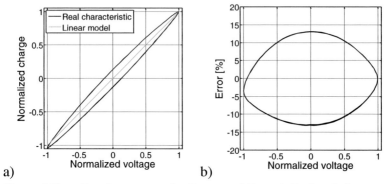

*Figure 3.13 – Capacitance error by linear modeling. a) Real and linear model characteristics, b) capacitance modeling error.*

The reconstructed transfer function defined in equation (3.53) is written as a function of the linear modeling error $\tilde{C}_p$. Considering that the hysteretic modeling error $\tilde{C}_p[\cdot]$ of equation (3.54) can not be directly used in equation (3.53), one can assume at a first approximation to use its maximum value (which typically corresponds to the passing through zero of the capacitance characteristic) to evaluate the quantity of the error introduced in the reconstruction, and consequently evaluating the moving of the zeros.

In order to reduce the reconstruction error, especially at high driving voltages where the hysteretic behaviour becomes consistent, a hysteretic reconstructor is suggested in this work, where the electrical clamped capacitance is modeled by a hysteretic model [GTJ13]. The hysteretic reconstructor is then defined as follows:

$$R(v,q) = q - \hat{C}_p[v],\qquad(3.55)$$

and the capacitance modeling error is defined as:

$$\tilde{C}_p[\cdot] = \hat{C}_p[\cdot] - C_p[\cdot].\qquad(3.56)$$

Such reconstructor aims at modeling the full nonlinear behaviour of the clamped electrical capacitance. Obviously, if it could be directly measured at all driving voltages and frequencies of interest, the modeling error would resemble uniquely the mathematical approximation of the chosen mathematical hysteretic model. Thus the modeling error could drop down to 1% or 2% in most cases. Unfortunately, it is not possible to measure directly the needed quantities, i.e. voltage and electrical charge. In fact, in order to bring the mechanical charge to zero, the piezoelectric material should be bonded on a structure with infinite stiffness and mass. Therefore, in this section, an identification method of the electrical clamped capacitance that minimizes the reconstruction error is presented.

As shown in section 3.3.2, the mechanically induced charge of a piezoelectric actuator bonded on a PVS is uncorrelated to the input driving voltage when this is white noise, or when it is 90° shifted respect to the mechanical charge. Nevertheless, such conditions do not hold in the case of hysteretic capacitance identification. In fact, white noise can not excite all amplitudes of the hysteretic capacitance and a 90° shift between mechanical charge and driving voltage does not lead to the necessary uncorrelation in the case of a hysteretic model. Consequently other approaches must be found.

Linear PVSs are characterized by resonances and antiresonances. The latter in particular approximates the favourable condition of high stiffness and very low deformation, corresponding to very high mechanical impedances, making them useful to be used for the identification data collection. A way to minimize the modeling error is, in fact, identifying the voltage-charge characteristic at an antiresonance frequency where it resembles closely the real clamped electrical capacitance. In fact, the measured charge will be the sum of the electrical charge and the mechanical charge, which is very low and shifted of 90° respect to the driving voltage at antiresonance. In particular, when the mechanical charge is shifted of ±90° with respect to the driving voltage, the voltage-charge peak-to-peak amplitude is equal to the one of the electrical clamped capacitance. It does instead affect the curvature of the characteristic, and consequently the phase shift of the reconstructed signal. These two conditions allow the identification of the electrical clamped capacitance affected by a low systematic error, due to the residual mechanical charge, in the frequency range around the chosen antiresonant frequency. In force of this approach, one can state that the electrical charge is approximately the total charge at antiresonant frequency condition:

$$q_e \approx q|_{f=f_{ar}} ,$$

(3.57)

where $f_{ar}$ is an antiresonant frequency of the structure.

The identification signal has then to fulfill two properties:

- Persistently exciting for the chosen hysteretic model
- Its spectrum must consist of an harmonic pulse located at an antiresonant frequency of the structure

The condition of persistent excitation depends on the hysteretic model chosen to identify the hysteretic behaviour. In this work, the Modified Prandtl-Ishlinskii (MPI) model has been chosen, and more details can be found in Appendix A. In such case, the identification signal consists of a sinusoidal signal whose frequency is tuned at the chosen antiresonance and its amplitude excites different voltage amplitudes within the chosen amplitude working range. This identification is performed offline after identifying the antiresonant frequencies of the structure in the working frequency range. This last step is very important, since a wrong identification of the antiresonances could lead to consistent errors of the reconstructed signal. Thus, either a very accurate model of the structure is provided or some other identification routines have to be used in order to locate the antiresonances.

Figure 3.14(a) shows a simulation result, where the real electrical capacitance characteristic is shown in black and the results obtained with a MPI model in blue. The charge used for the identification of the model has been measured in antiresonance where there is a -90° shift of the mechanically induced charge respect to the electrically induced charge. Moreover, a 5% of the overall charge has been attributed to the mechanically induced charge, which can be considered a relatively common percentage in most of piezoelectric structures. In Figure 3.14(b), instead, the error is shown still within the 5% of mechanically induced charge but at different phase shifts.

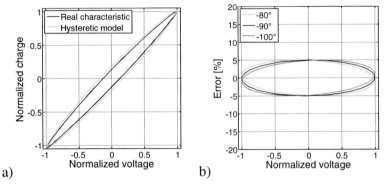

*Figure 3.14 – Capacitance error by hysteretic modeling. a) Real and hysteretic model characteristics, b) modeling error with the mechanically induced charge at different phase shifts respect to the electrically induced charge.*

As one can see, a -90° shift will result in zero error at the extremes of the characteristic, and with the mechanical charge error present at the driving voltage crossing zero. Little devations from the -90° shift can lead to a small mismatch in the peak-to-peak amplitude reconstruction, which can be neglected. Thus, even in case the antiresonances can not be perfectly matched, it is still possible to perform a good reconstruction, since these errors are still much lower than the maximum error obtained by using a linear reconstruction, and consequently the quantity $\tilde{C}_p$ in equation (3.53) is significantly reduced. In fact, since the curvature of the hysteretic characteristic is better matched, one will get a lower error in the phase shift reconstruction.

Moreover, considering the typical creep effect of piezoelectric materials, one must also consider that such technique performs better in the frequency range around the chosen antiresonant frequency, since far from this frequency the hysteretic electrical clamped capacitance would differ due to nonlinear time dependent effects, as shown in section 2.1.3. Thus, a hysteretic capacitance must be identified per each narrow frequency range of interest, and each identified model can be used for self sensing reconstruction by properly filtering the measured quantities. Nevertheless, since the rate of change of the capacitance due to creep is much higher at low frequencies than at high ones, the necessity of identifying more characteristics is higher at low frequencies (see Figure 2.7). Such self sensing reconstruction has been implemented digitally in this work. In fact, a hysteretic model is more easily implementable on a microcontroller rather than by using analogic circuitries.

## 3.6. Self sensing techniques implementation

All self sensing techniques have in common the measurement of voltage and charge of the piezoelectric actuator. Based on such information, the self sensing reconstruction can be implemented analogically, digitally or with a hybrid configuration.

### 3.6.1. Self sensing analogic implementation

The bridge circuit shown in Figure 3.7 realizes an analogic reconstruction.

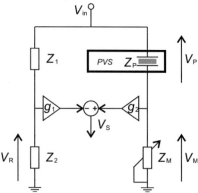

*Figure 3.15 – A self sensing impedance bridge circuit for strain measurement with variable reference impedance $Z_M$.*

While the right branch provides a measurement of the charge on the piezoelectric actuator, $V_M$, the left branch provides a measurement of the voltage, $V_R$. In order to balance the bridge, the gains $g_1$ and $g_2$ are implemented. These gain stages are usually realized by using standard operational amplifiers based circuitries, and their gains can be easily tuned by using potentiometers.

Another way to balance the bridge is by adjusting the measuring impedance $Z_M$ as shown in Figure 3.15. This impedance is generally made of the parallel of a capacitor $C_M$ and a resistor $R_M$ (see Figure 3.17). Varying the value of a resistor can be easily done by using a normal potentiometer, while varying the value of a capacitor requires particular electronics. At this aim, a capacitance multiplier can be implemented in place of a typical capacitor. Such circuitry allows to simulate a variable capacitor whose value can be changed by acting on some potentiometers. More details can be found in section 4.1.

The main advantage of using an analogic circuitry for self sensing reconstruction is represented by the wider frequency range of operation, since no antialiasing filters are needed nor limiting sampling frequencies. The only significant delays come from the operational amplifiers, which can insert a response delay of approximately 1 μs each. Thus such circuitries can perform perfectly up to several hundreds kHz, without introducing significant phase delays. Moreover, the measurement voltage is produced analogically and consequently are not affected by discretization which is typical of digital to analog converters (DACs) and degrade the quality of reconstructed signal.

Nevertheless, these circuits have to be tuned manually by acting on the potentiometers of the gains or of the reference impedance, making the usage of these circuits mostly useful for experimental applications or quick investigations.

### 3.6.2. Self sensing digital implementation

A way to overcome the main drawback of the analogic implementation, which is the manual tuning of the bridge, is the digital implementation of the self sensing reconstruction. Moreover, vibration controllers are often implemented on a microcontroller, thus additional hardware would not be required. The typical scheme of digital self sensing reconstruction is shown in Figure 3.16.

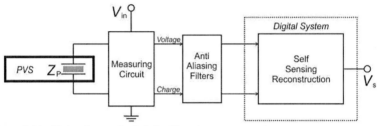

*Figure 3.16 – Main scheme of a digital self sensing reconstruction system.*

The piezoelectric actuator $Z_p$ is connected to a measuring circuit, which measures voltage and charge while allowing the piezoelectric material to be driven with the voltage $V_{in}$. Voltage and charge measurements are then treated by the self sensing reconstruction unit executed on the digital system after being filtered by the anti aliasing filters.

The measuring circuit is the core of the self sensing reconstruction, since it measures simultaneously the voltage and the charge of the piezoelectric actuator. A typical circuit is called Sawyer Tower circuit [ST30], a name which comes from the two scientists who proposed it in 1930. Sawyer and Tower, in fact, proposed such a circuit to measure the polarization hysteresis cycle in function of the electric field strength of a ferroelectric material. The electric schematic of such circuit is shown in Figure 3.17.

The Sawyer Tower circuit is practically an impedance voltage divider, where the output voltage $V_M$ is proportional to the charge stored on the piezoelectric actuator. Moreover, a resistive voltage divider is used to provide the voltage $V_p$ that is proportional to the voltage driven by the amplifier, $V_{in}$. More details about its properties and dynamic behaviour can be found in section 4.2.

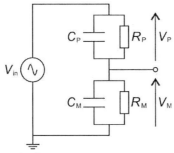

*Figure 3.17 – Sawyer Tower circuit.*

It has to be noticed that the part of the input voltage $V_{in}$ dropping on the piezoelectric actuator depends strongly by the value of the capacitance $C_M$. Thus this value is chosen to be around 100 times the value of the piezoelectric capacitance $C_P$, which allows around the 99% of the input voltage $V_{in}$ to drop over the piezeoelectric transducer. Nevertheless, this choice can be a drawback in some cases since it can reduce the amplitude of the measurement voltage $V_M$. For such reason, an active version of such circuit is here proposed, named Active Sawyer Tower, whose main schematic is shown in Figure 3.18 (more details can be found in section 4.3).

Such circuitry separates the piezoelectric impedance $Z_P$ from the measuring impedance $Z_M$ by using an operation amplifier. In this case the full driving voltage $V_{in}$ drops over the piezoelectric actuator, while the dimensioning of the measuring impedance is only limited by the power of the chosen operation amplifier. Again, a resistive voltage divider can be used to measure the input voltage.

The measurement signals provided by the measuring circuit need to be conditioned before being processed by the digital system. In fact, they first need to be low-pass filtered by the anti aliasing filters, which are tuned on the chosen sampling frequency, and then are converted by analog to digital converters (ADCs). Generally, the dimensioning of such circuits is made in a way that the amplitude of the signals can cover the whole input voltage range of the ADCs in order to maximize the signal resolution. Thus, the Active Sawyer Tower is more suitable.

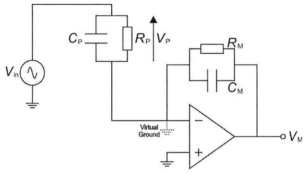

*Figure 3.18 – Active Sawyer Tower circuit.*

The digital self sensing reconstruction algorithm running in the digital system has the task to solve the reconstruction. The reconstruction unit can be either linear, as shown in equation (3.50), or hysteretic, as shown in equation (3.55). The electric displacement and the electric field strength are then proportional to the measured signals, $V_M$ and $V_P$, according to the parameters of the measuring unit and eventual gains in the filtering stage.

This kind of implementation has particular advantages. In fact, not only the reconstructor unit parameters can be changed rather quickly, and during execution, but it allows a quick implementation of complex reconstruction algorithms, like the hysteretic reconstruction that makes use of a hysteretic mathematical model, which is typically characterized by a high number of parameters and nonlinear functions. The major drawback, instead, lies in the limited bandwith which is due both to the sampling frequency and to the low-pass filtering of the anti aliasing filters. Moreover, the sensing signal provided by the reconstruction is affected by discretization due to the digital to analog converters (DACs).

### 3.6.3. Self sensing digital and hybrid implementations with adaptive identification

The use of a LMS algorithm for identifying the electrical clamped capacitance by driving the piezoelectric material with white noise as described in section 3.3.2 is quite effective and can be used for tuning the bridge circuit. Nevertheless, the piezoelectric capacitance can change over the time due to aging or temperature. Thus an online identification would be more effective since able to track such variation during operation. In particular, the online identification has to be performed while driving the piezoelectric actuator. The idea is that a low voltage noise can be added to the driving voltage in a frequency range out of the frequency range of operation and then isolated for identification by properly designed filters.

Thus, the identification signal is a colored noise, whose band width is limited by the high pass filters necessary to select the identification information and, when using a digital system, by the antialiasing filters. Such filtering process increases the correlation among the samples of the noise signal, leading to a bias error (second term of equation (3.34) is not zero). Nevertheless, it is experimentally observed that the bias error induced by filtering is very small and can often be neglected. It is also important that the harmonic driving voltage for operation does not exceed the cut-off frequency of the high pass filters, or the bias error would increase.

A scheme of the digital self sensing reconstruction with adaptive identification is shown in Figure 3.19.

A hybrid implementation of the self sensing reconstruction with adaptive identification, instead, can join the advantages of the analogic reconstruction, i.e. a wide frequency band of operation, and the digital identification, i.e. the capability of solving the LMS algorithm for identifying the electrical clamped capacitance. A scheme of the hybrid implementation is shown in Figure 3.20.

In this case, the self sensing reconstruction is performed by a tunable bridge circuit, which consists of a bridge circuit, as the one shown in Figure 3.7, where the gain $g_2$ is kept constant, and the gain $g_1$ can vary. In this way the self sensing reconstruction can be performed over a wide frequency band, without particular phase delays nor any kind of discretization. A useful component to realize a tunable bridge circuit is the analog multiplier, as for example the AD632 from Analog Devices [VC96]. The gain $g_1$ is then driven by the adaptive filter which is executed by the digital system. In fact, the adaptive filter is easily implementable on a microcontroller, while an analogic realization would require particular effort. Colored noise is added to the harmonic driving signal for identification purposes, both in the case of the digital and hybrid implementation.

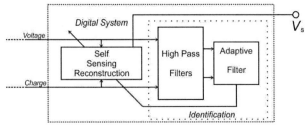

*Figure 3.19 – Main scheme of a digital self sensing reconstruction system with adaptive identification.*

Unfortunately, the usage of a hybrid implementation in the case of the hysteretic reconstruction is not easily realizable. In fact, this would require an analogic circuit capable of simulating a hysteretic capacitance, with tunable parameters. The design of such kind of circuitry could be particularly complex and effort demanding, and will not be discussed in this work.

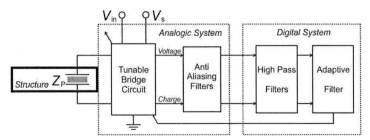

*Figure 3.20 – Main scheme of a hybrid self sensing reconstruction system with adaptive identification.*

# 4.  Electronics for self sensing applications

Electronics used for self sensing applications has to be particularly designed in order to provide reliable measurements for the self sensing reconstruction. This chapter describes four electronic circuits.

The first circuit in section 4.1 is a variable capacitance multiplier, which simulates a variable capacitor, and it is very useful when implementing a linear self sensing reconstruction by means of analogic circuits.

The second and third circuits are respectively the Sawyer Tower (section 4.2) and the Active Sawyer Tower (section 4.3), which allow the measurement of the charge stored on a piezoelectric actuator while driving it. As discussed in Chapter 3, such circuits are the main core of the self sensing digital implementation. Their tuning and frequency responses are here shown and discussed.

Finally, an Active Sawyer Tower circuit with temperature measurement is proposed in section 4.4. Such circuit in fact allows to measure simultaneously voltage, charge and electrode temperature by using a third wire connection and the Seebeck effect. It is useful for those applications where temperature variation can not be neglected.

## 4.1.  Variable capacitance multiplier

A variable capacitance multiplier allows to simulate a variable capacitor by acting on a potentiometer. The principle schematic is shown in Figure 4.1.

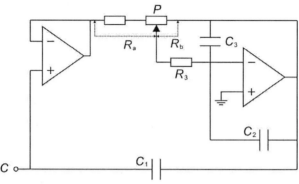

*Figure 4.1 – Variable capacitance multiplier principle schematic.*

The circuit is based on active components, i.e. the two operational amplifiers. In particular, the simulated capacitance seen at the pin $C$ respect to ground can be calculated via the following formula:

$$C = C_1 \left( 1 + \frac{R_b}{R_a} \right),$$

(4.1)

where $R_a$ and $R_b$ are varied by means of the potentiometer $P$. This circuit is then able to multiply the value of the capacitance $C_1$, which is the reference capacitance for this circuit.

A capacitance multiplier of this kind can be integrated in the bridge circuit of Figure 3.15 in place of the measuring impedance $Z_M$. In this way it is easily possible to tune the bridge circuit for self sensing reconstruction without acting on the gains $g_1$ and $g_2$. Moreover, by using a digital potentiometer for adjusting the multiplied capacitance, a digital system could be used to tune the reconstruction bridge circuit adaptively. Similarly to what described in section 3.6.3, a hybrid implementation of the self sensing reconstruction can be realized by digitally adjusting the measuring impedance $Z_M$.

## 4.2. Sawyer Tower circuit

The Sawyer Tower circuit is a passive circuit which allows to measure simultaneously the voltage and the charge stored on a piezoelectric transducer. The conceptual schema is shown in Figure 4.2.

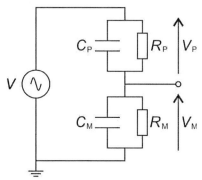

*Figure 4.2 – Sawyer Tower circuit principle schematic.*

The idea is to place a measuring impedance in series with the piezoelectric element, which is composed of a capacitor $C_M$ and a resistor $R_M$. These are respectively called measuring capacitor and measuring resistor since the voltage across the capacitor is proportional to the charge present on the piezoelectric material at high frequencies. Moreover, the resistor $R_M$ is needed in order to drain the leakage current of the piezoelectric actuator.

The transfer function between the measured voltage $V_M$ and the applied voltage $V$ of this board is shown in the following equation:

$$\frac{V_M(s)}{V(s)} = \frac{R_M}{R_M + R_P} \frac{1 + R_P C_P s}{1 + \dfrac{(C_P + C_M) R_P R_M}{R_M + R_P} s}. \qquad (4.2)$$

This is a first order proper transfer function, which is characterized by the frequencies of the zero and the pole, respectively $f_Z$ and $f_P$ (the latter one also called cut-off frequency) which can be expressed as:

$$f_Z = \frac{1}{2\pi R_P C_P},$$
(4.3)

$$f_P = \frac{R_M + R_P}{2\pi R_P R_M \left( C_M + C_P \right)}.$$
(4.4)

Moreover, the frequency $f_Z$ depends only on the piezoelectric actuator impedance while $f_P$ depends also on the measuring impedance.

In order to dimension the Sawyer-Tower circuit, it is important to consider the static gain $K_0$ and the gain at high frequencies $K_{hf}$, which result to be:

$$K_0 = \frac{R_M}{R_M + R_P},$$
(4.5)

$$K_{hf} = \frac{C_P}{C_M + C_P}.$$
(4.6)

The resistor $R_M$ has an effect simultaneously on the static gain and on the cut-off frequency. In particular, the latter has a bigger importance in determining the correct resistance since the measurement offset (or static gain) can be measured and compensated by using simple circuitry when not wanted.

The capacitance $C_M$, instead, has an influence on more aspects. Primarily, it determines the high frequency gain $K_{hf}$. By increasing the measuring capacitance, the gain becomes smaller, having consequently a loss in measurement resolution, even though the amount of voltage that drops over the active element becomes bigger. Considering that the voltage on the piezoelectric material $V_P$ is given by the following equation

$$V_P = \frac{C_M}{C_M + C_P} V,$$
(4.7)

it is a common choice to set $C_M = 100C_P$ so that the voltage on the active material is equal to 99.01% of the driving voltage while still having an acceptable resolution.

| Parameter | Value |
|---|---|
| Measuring Capacitance $C_M$ | 100 μF |
| Measuring Resistance $R_M$ | 1 MΩ |
| Piezoelectric Capacitance $C_P$ | 1 μF |
| Leakage Resistance $R_P$ | 400 MΩ |

*Table 4.1 – Parameters of the simulated Sawyer Tower.*

By considering the values shown in Table 4.1, the two frequencies $f_Z$ and $f_P$ and the two gains $K_0$ and $K_{hf}$ can be calculated, and they are listed in Table 4.2.

| Parameter | Value |
|---|---|
| $f_Z$ | 0.39 mHz |
| $f_P$ | 1.6 mHz |
| $K_0$ | −52.06 dB |
| $K_{hf}$ | −40.08 dB |

*Table 4.2 – Zero and pole cut-off frequencies, static and high frequency gains of the simulated Sawyer Tower.*

The obtained Bode diagram for such Sawyer Tower is shown in Figure 4.3.

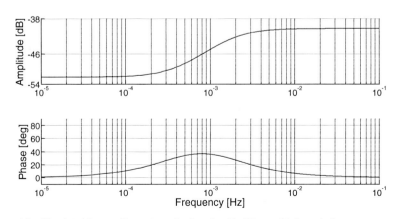

*Figure 4.3 – Simulated Sawyer Tower transfer function $V_M / V$: amplitude and phase.*

## 4.3. Active Sawyer Tower circuit

An alternative circuit for measuring the charge while driving the piezoelectric actuator is the so-called Active Sawyer-Tower, whose conceptual schema is shown in Figure 4.4.

This circuit makes use of an operational amplifier to create a virtual ground point for the piezoelectric element while measuring the charge flowing through it. In this case the measuring impedance is placed in the feedback path of the operation amplifier, which will be driven in order to keep the voltage at its negative input equal to ground, thus creating a virtual ground.

By applying the Kirchhoff's current law on the negative input of the operational amplifier, one can get the transfer function which is shown in the following equation:

$$\frac{V_M(s)}{V(s)} = -\frac{R_M}{R_P}\frac{1+R_PC_Ps}{1+R_MC_Ms},\qquad(4.8)$$

and has a similar structure of the one presented in equation (4.2) for the Sawyer-Tower circuit. Also in this case, in fact, it is characterized by a zero and a pole, whose frequencies $f_Z$ and $f_P$ can be expressed as:

$$f_Z = \frac{1}{2\pi R_P C_P},\qquad(4.9)$$

$$f_P = \frac{1}{2\pi R_M C_M}.\qquad(4.10)$$

In this case, the cut-off frequency $f_P$ is a function only of the measuring impedance and does not depend on the piezoelectric element, since the infinite input impedance of the operational amplifier splits the circuit in two parts, namely the piezoelectric circuit and the measuring one. In this case the static gain $K_0$ and the high frequency gain $K_{hf}$ are expressed as:

$$K_0 = -\frac{R_M}{R_P},$$ (4.11)

$$K_{hf} = -\frac{C_P}{C_M}.$$ (4.12)

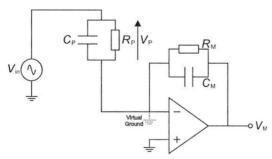

**Figure 4.4 – Active Sawyer Tower circuit principle schematic.**

Again, the measuring capacitance and the measuring resistance are determined in order to shape the frequency behaviour of the circuit in a similar way of the case of the Sawyer-Tower circuit.

By considering the values shown in Table 4.3, the zero and pole frequencies as well as the static gain and the gain at high frequencies are calculated and listed in Table 4.4.

| Parameter | Value |
|---|---|
| Measuring Capacitance $C_M$ | 100 μF |
| Measuring Resistance $R_M$ | 1 MΩ |
| Piezoelectric Capacitance $C_P$ | 1 μF |
| Leakage Resistance $R_P$ | 400 MΩ |

**Table 4.3 – Parameters of the simulated Active Sawyer Tower.**

| Parameter | Value |
|---|---|
| $f_z$ | 0.39 mHz |
| $f_P$ | 8 mHz |
| $K_0$ | −52.04 dB |
| $K_{hf}$ | −26.02 dB |

**Table 4.4 – Zero and pole cut-off frequencies, static and high frequency gains of the simulated Active Sawyer Tower.**

The Bode diagram of the simulated Active Sawyer Tower is shown in Figure 4.5.

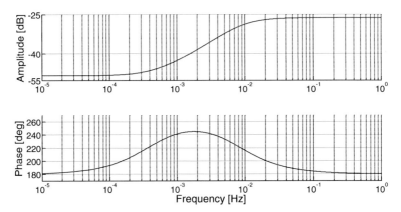

*Figure 4.5 – Simulated Active Sawyer Tower transfer function $V_M / V$: amplitude and phase.*

The main difference between the classic Sawyer Tower and the Active Sawyer Tower circuits is given by the feature of the Active Sawyer Tower of separating the piezoelectric element by the measuring impedance via the infinite input impedance of the operational amplifier. Two relevant effects can be considered. First, the driving voltage $V_{in}$ drops totally on the piezoelectric material ($V_P = V_{in}$) and consequently we can use the full voltage range of the amplifier. Second, the choice of the measuring impedance modifies only the pole of the transfer function of equation (4.8) without affecting the voltage on the piezoelectric material. Moreover, the Active Sawyer Tower can work with different piezoelectric materials without modifying its frequency behaviour, despite of the classic Sawyer Tower whose pole dynamic is influenced by the piezoelectric capacitance and resistance. Indeed, this is a very important aspect, since typically a piezoelectric actuator changes its capacitance according to the driving voltage amplitude and frequency and consequently shifting the pole frequency.

The piezoelectric values shown in Table 4.1 and in Table 4.3 refer to a typical piezoelectric multilayer actuator which typical driving voltage is equal to ±75 V. In the case of the Sawyer Tower circuit, and considering a measuring impedance equal to 100 times the piezoelectric one for the reasons discussed above, one would get a maximum measured voltage $V_M$ equal to 0.7426 V, which does not lead to a high resolution when discretized by the DAC converters. In the case of the Active Sawyer Tower instead, the maximum measured voltage becomes 3.75 V by choosing a 20 µF as measuring capacitance, despite an increase of the cut-off frequency from 1.6 mHz to 8 mHz, which for the majority of applications is neglectable. Theoretically, as long as one can neglect the cut-off frequency since it is very low with respect to the typical applications, one could choose lower values of the measuring capacitance in order to increase the measurement resolution.

## 4.3.1. Active Sawyer Tower for power demanding applications

The Active Sawyer Tower allows a more flexible dimensioning with respect to the Sawyer Tower circuit by decoupling the piezoelectric actuator impedance from the measuring impedance via an operation amplifier. Nevertheless, in order to realize this circuit it is important to consider the limitations of a real operational amplifier and the way they affect the functioning of this circuit. In particular, the ideal operational amplifier is characterized by an infinite input impedance, a zero output impedance and an infinite gain-bandwidth product while real operational amplifiers have limited performances. Consequently it is of interest to analyze how these limitations affect the functioning of the Active Sawyer Tower.

The input resistance of a real operational amplifier is usually a value greater than 1 MΩ and can get up to $10^{12}$ Ω for the ones based on a JFET input stage while the input capacitance is usually on the order of the pF. Consequently, it has to be taken into account that a part of the current flowing into the piezoelectric material is flowing through the input impedance of the operational amplifier, determining a measurement error, which is as smaller as higher the input impedance. Nevertheless the value of the input impedance is normally so high that one can neglect the amount of charge flowing between the two inputs of the operational amplifier, considered also that the voltage difference is very low.

The non-zero output impedance of the real operational amplifier, instead, limits its power, since the output current dissipates heat while flowing through it. Consequently, each operational amplifier has a limited output current in order not to damage the device itself and therefore has a power limitation. These two parameters are fundamental for choosing the correct operational amplifier for the Active Sawyer Tower, i.e. current and power. In particular, considering a capacitance $C$, which is driven with a sinusoidal voltage $V(t) = V_{\text{MAX}} \sin(2\pi f t)$, the maximum current is:

$$i_{\text{MAX}} = 2\pi f C V_{\text{MAX}},\tag{4.13}$$

and the maximum power peak required to drive the capacitance is equal to:

$$P_{\text{MAX}} = \pi f C V_{\text{MAX}}^2.\tag{4.14}$$

Considering equation (4.14), the maximum reactive power peak $P_A$ of the voltage amplifier driving the piezoelectric actuator and the maximum reactive power peak $P_{\text{AST}}$ required to the operational amplifier to drive the measuring capacitance are:

$$P_A = \pi f C_P V^2,\tag{4.15}$$

$$P_{\text{AST}} = \pi f C_M V_M^2.\tag{4.16}$$

It can then be shown by means of the equation (4.12) that there exist a relation between $P_A$ and $P_{\text{AST}}$ that is expressed by the following equation:

$$P_{\text{AST}} = \frac{C_P}{C_M} P_A.\tag{4.17}$$

Thus, the measuring capacitance determines not only the gain at high frequency of the circuit and the position of the cut-off frequency, but also the maximum power peak required to the operational amplifier to perform the measurement.

Using the values shown in Table 4.3 and considering a driving voltage of ±75 V at a frequency of 1 kHz, which represents a typical application, the maximum power peak required to the voltage amplifier is equal to 17.67 var while the maximum power peak required to the operational amplifier is 0.88 var. Thus in many applications power electronics is required and power losses have to be taken into consideration for dimensioning appropriate cooling systems. In fact, the mean power losses for a bipolar amplifier are expressed as:

$$P_{bp} = fC\frac{V_n V_{pp}}{2},$$
(4.18)

where $V_n$ is the nominal peak-to-peak voltage and $V_{pp}$ is the capacitor driving peak-to-peak voltage. Consequently, the power losses of the operational amplifier are equal to a maximum of 2.25 W, considered a nominal voltage $V_n$ of 30 V (±15 V).

Moreover, the limited gain-bandwidth product limits the performances over the frequency. This value is usually on the order of some MHz and can be neglected in most of the self sensing applications where the maximum operating frequency is on the order of some kHz. Nevertheless, this effect is related also to the limited slew rate of the operational amplifier, which introduces a small delay in the measurement. This latter limitation has an evident effect on the functioning of the circuit. In fact, the delay due to the operational amplifier will produce a non-zero voltage at the virtual ground point. Thus, the real circuit of the Active Sawyer Tower is shown in Figure 4.6.

In this circuit, the capacitor $C_{COLL}$ has been introduced between the virtual ground point and the real ground, which is called collector capacitor, in order to collect the charge that the operational amplifier can not provide at its output, which is fundamental for the stability of the circuit. The virtual ground voltage is an outstanding parameter of the circuit since it is an indicator of the performance of the operational amplifier.

The limitations shown above have then to be considered in the choice of the operational amplifier. In particular, the power requirement represents the most important aspect since usual operational amplifiers are made for low power applications. For power demanding applications two different implementations of the Active Sawyer Tower are possible. The first solution is based on a low voltage high current operational amplifier, like for example the OPA548 produced by Texas Instruments. This operational amplifier is able to provide a maximum continuous current of 3 A and a current peak of 5 A, with a voltage output swinging between -26.7 V and 26.3 V. Its main parameters are shown in Table 4.5.

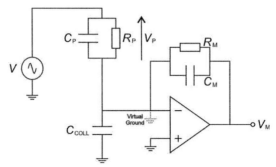

*Figure 4.6 – Active Sawyer Tower circuit real schematic.*

| Parameter | Value |
|---|---|
| Input Resistance | 10 MΩ |
| Gain-Bandwith Product (GBW) | 1 MHz |
| Slew Rate | 10 V/μs |
| Output Current Limit Range | ±5 A |

*Table 4.5 – Parameters of a OPA548 from Texas Instruments.*

The second implementation of the Active Sawyer Tower is based instead on a low power operational amplifier, in this case a TL071 produced by ST Microelectronics, whose performances are enhanced by means of power transistors. The parameters of this operational amplifier are shown in Table 4.6.

In this work, the operational amplifier TL071 is used because of its high input resistance and its high bandwidth, while the task of sourcing or sinking the needed current is left up to power transistors. A conceptual schematic of a current boosting stage is shown in Figure 4.7.

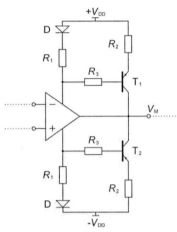

*Figure 4.7 – Low power operational amplifier boosted by using power transistors.*

In Figure 4.7 the transistors $T_1$ and $T_2$, respectively PNP and NPN BJT transistors, are used in order to source and sink the current of the load. They are driven via the sensor current represented by the resistors $R_1$, with a gain due to the ratio $R_1/R_2$ and whose electrical working point is determined via the diodes D. The resistors $R_3$ are instead used to limit the current at the base of the transistors. This configuration is then expandable by adding other power stages at the output of the operational amplifier. In this way, one can reach the required maximum current without changing the performances of the operational amplifier. Figure 4.8 shows an Active Sawyer Tower with 8 stages of current boosting realized at the LPA department of the Saarland University.

| Parameter | Value |
|---|---|
| Input Resistance | 1 TΩ |
| Gain-Bandwith Product (GBW) | 4 MHz |
| Slew Rate | 16 V/μs |
| Output Current Limit Range | ±40 mA |

*Table 4.6 – Parameters of a TL071 from ST Microelectronics.*

Using a power operational amplifier has the advantage of having a simpler design due to the lower number of components, which results in a smaller circuit and an easier cooling system design. At the same time the availability of such components does not always match the requirements of the particular application, and it is usually more expensive. On the other side realizing a power component by boosting a low operational amplifier via discrete power components presents the advantage of achieving the application specifications considered the modularity of the circuit, which can be enhanced by adding more power output stages.

*Figure 4.8 –Active Sawyer Tower with 3 stages of current boosting realized at the LPA department of the Saarland University.*

This realization results cheaper than the previous one despite an increase in size and complexity of the cooling system which has to handle the power losses of all the used transistors. The cooling system adopted in the Active Sawyer Tower of Figure 4.8 is made of a passive heat sink combined with a fan in order to increase the cooling performance.

## 4.4. Active Sawyer Tower with temperature measurement

As shown in the constitutive equations (2.1) of a piezoelectric material, temperature affects the permittivity (and therefore the capacitance) of a piezoelectric material. While in some self sensing applications this effect could be neglected, as in the case of a piezoelectric plate bonded on an aluminium vibrating structure which behaves as a heat sink, in some other applications temperature can not be neglected.

A temperature change can occur either because of a change of the environmental temperature, or due to self heating, as discussed in section 2.1.3. Nevertheless, while the room temperature can be easily measured with a thermometer, the temperature of the piezoelectric ceramic is not directly accessible. Common temperature sensors need to be bonded on the piezoelectric material and this could lead to unexpected results. In fact, the piezoelectric elongation due to the applied voltage generates heat due to the friction between the temperature sensor and the piezoelectric material itself.

An improved version of the Active Sawyer Tower circuit is here proposed. This circuit is able to measure simultaneously voltage, charge and temperature of the piezoelectric material for dynamic driving operation, by soldering a third wire on the piezoelectric electrode. This wire is made of constantan and it is soldered on the negative electrode, where the other wire, made of copper, is usually soldered for driving the piezoelectric material. These two wires create a type-T thermocouple which can be used for measuring the temperature of the negative electrode. This approach is based on the idea that the electrode temperature is equal to the ceramic temperature, since the electrode is usually spread over the whole surface of the ceramic. This hypothesis is even more valid for multilayer actuator, where the negative electrode extends within the ceramic block.

A principle schematic of the Active Sawyer Tower with temperature measurement is shown in Figure 4.9.

As one can see, two wires are soldered at the negative electrode, coloured in blue and orange, which are respectively made of constantan and copper. They are both connected to a Cold Junction Compensation (CJC) unit: the difference in temperature $\Delta T$ between the soldering point on the negative electrode and the room temperature will result in a very small voltage (48.2

μV/°C) at the input pins of the CJC unit due to the Seebeck effect. The CJC unit provides a voltage proportional to the temperature of the negative electrode by measuring and compensating the room temperature. Since the sensitivity of this thermocouple system is very low, the CJC unit performs a high gain amplification (usually between 200 and 1000).

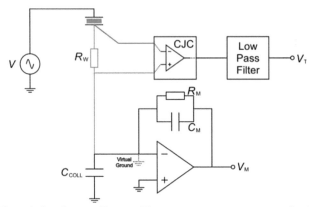

*Figure 4.9 – Active Sawyer Tower with temperature measurement circuit principle schematic. Orange: Copper – Blue: Constantan.*

It is also very important to dimension properly the copper wire. In fact, the current flowing through the piezoelectric material flows only in the copper wire, since the CJC unit has a very high input impedance and absorbs nearly no current. Consequently a voltage drop on the copper wire is present, and the CJC unit amplifies the sum of the Seebeck voltage and the voltage drop on the copper wire, which depends on its resistance $R_W$. The voltage provided by the CJC unit will then consist of two parts: the temperature voltage and a voltage proportional to the current flowing through the piezoelectric material. These two quantities can be decoupled by assuming that the piezoelectric material is driven dynamically in a specified frequency range. In fact, since thermal behaviours are typically very slow, and piezoelectric materials are typically used for dynamic operation, it is often possible to filter out the current components from the voltage provided by the CJC unit.

A prototype of this circuit has been realized and tested on a multilayer actuator, which is a PICMA P-885.10 produced by PI Ceramic, whose parameters are listed in Table 4.7.

The CJC unit is a AD849x chipset from Analog Devices, while the low pass filter is a 10th order digital filter whose cut-off frequency is at 40 Hz, implemented on a dSpace DS1103. Moreover, the Active Sawyer Tower measuring capacitance $C_M$ and measuring resistance $R_M$ are respectively 20 µF and 10 MΩ.

| Parameter | Value |
|---|---|
| Dimensions | 5x5x9 mm³ |
| Maximum Voltage | 120 V |
| Maximum Displacement at 120 V | 8 µm |
| Blocking Force at 120 V | 800 N |
| Stiffness | 100 N/µm |
| Electrical Capacitance | 0.68 µF |

*Table 4.7 – Parameters of the PICMA P-885.10 produced by PI Ceramic.*

The piezoelectric actuator has been hanging unconstrained in air and driven with a sinusoidal voltage of 90 V amplitude and 50 V offset at 400 Hz. The driving voltage induces a temperature change due to self heating. The temperature behaviour measured by the circuit is shown in Figure 4.10.

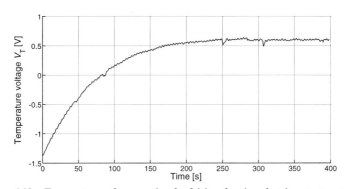

*Figure 4.10 – Temperature voltage vs. time by driving the piezoelectric actuator at 400 Hz, 90 V amplitude and 50 V offset.*

The scalar factor between the measured temperature voltage $V_T$ and the real temperature depends on the parameters of the CJC unit, which have not been identified in this work. Nevertheless the temperature voltage $V_T$ that corresponds to room temperature is equal to -1.39 V.

As one can see, the temperature behaviour resembles the step response of a first order system as it is expected. Figure 4.11a shows the hysteretic piezoelectric unconstrained capacitances measured at 4 time instants (thus at 4 different temperatures) over the 400 s of observation of the experiment shown in Figure 4.10, respectively at 0 s, 30 s, 90 s and 400 s.

The peak to peak amplitude of the hysteretic capacitance increases over temperature, as expected. In fact, the peak to peak amplitude of the hysteretic capacitance increases of 11.7 % from the lowest temperature up to the highest one. Figure 4.11b shows similar results obtained by driving the piezoelectric transducer at 800 Hz in the same voltage range, which dissipates more power and thus leads to higher temperatures. In this case the hysteretic capacitance increases up to the 23.9% more than the one measured at room temperature.

This circuit can be used to track temperature variations of the piezoelectric material over time. A last experiment has been conducted by using an electric fan to cool down the piezoelectric actuator. The electric fan can be driven at low voltage (6 V) and at high voltage (12 V) to provide less or more air flow. The piezoelectric actuator, hanging unconstrained in the air, has been driven at 400 Hz to induce self heating at an amplitude peak to peak of 90 V with an offset of 50 V. Figure 4.12 shows the measured temperature behaviour over time.

This experiment is important to observe the repeatability of the temperature measurement. In fact, the measured temperature starts and ends at the room temperature (which is still -1.39 V as in the experiment of Figure 4.10) after the driving voltage is turned off, as expected. Futhermore, when the fan is off, the temperature settles always at the same value (0.52 V), which represents the maximum temperature reached by self heating.

It has to be remarked that the relation between the difference in temperature $\Delta T$ and the Seebeck voltage $V_{Sb}$ is nonlinear, and it can be expressed as:

$$\Delta T = \sum_{n=0}^{N} a_n V^n \,,$$

(4.19)

where the coefficients $a_n$ depend on the type of thermocouple. Anyways, the compensation of the nonlinearity has not been taken into consideration in this work since not of interest at the aim of verifying the proposed circuit. For this reason, the shown temperature behaviour is affected by uncompensated nonlinearities.

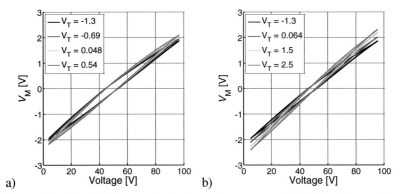

*Figure 4.11 –Piezoelectric unconstrained capacitance measured at different temperatures when a) driven at 400 Hz and b) driven at 800 Hz.*

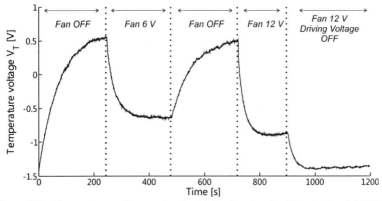

*Figure 4.12 – Temperature voltage vs. time by driving the piezoelectric actuator at 400 Hz at an amplitude of 90 V peak to peak and driving an electric fan for forced cooling.*

125

# 5. Experimental verification and results

This chapter aims at verifying experimentally the self sensing techniques proposed and discussed in the previous chapters, with the main focus of demonstrating that a hysteretic reconstructor, differently from the linear one, which is limited at low driving voltages, is able to perform a reliable reconstruction at all driving voltages.

In particular, the performance of a vibration controller that uses the reconstructed signal as feedback information has been compared when using a linear or a hysteretic reconstructor. Furthermore, in order to induce the vibration controller to drive the self sensing actuator at low and at high driving voltages (in order to excite the piezoelectric nonlinearities), a disturbance actuator has been driven to induce low as well as high level structural vibrations.

Section 5.1 shows the experimental results obtained by using a self sensing actuator bonded on an aluminum cantilevered beam. A linear and a hysteretic reconstructor have been implemented and their performance are compared by using a resonant controller designed for damping one structural resonant frequency.

Similar experiments have been conducted also on an partially clamped aluminum plate, and experimental results are shown and discussed in section 5.2. In this case the reconstruction performance are compared by means of two vibration controllers, which are a PPF controller and a resonant controller.

Finally section 5.3 summarizes the advantages and the drawbacks of both reconstructors confirmed by the experimental results.

## 5.1. *Piezoelectric cantilevered beam*

The first structure used for verifying the discussed self sensing techniques is a cantilevered beam, which is typically used for scientific investigations due to the simplicity of its implementation in a laboratory setup. More details have been presented in section 2.2.2.

In this experiment a piezoelectric actuator is driven to induce a disturbance vibration into the structure, and a self sensing actuator is then controlled to reduce the amount of structural vibration. In order to measure the level of the vibration, a third piezoelectric transducer, which serves as a sensor, is bonded coupled to the self sensing piezoelectric actuator.

A resonant controller is then implemented to compare the quality of the quantities reconstructed by a linear and a hysteretic reconstructor (see Appendix B).

### 5.1.1. Experimental set up description

The experimental set up consists of 5 parts:

- a cantilevered beam with three piezoelectric transducers bonded on it: the first one serves as disturbance actuator, the second one as a self sensing actuator and the third one as a piezoelectric sensor, which is coupled to the self sensing actuator,
- a measuring circuit for measuring simultaneously the voltage and the charge of the self sensing actuator,
- two voltage amplifiers, one for driving the disturbance actuator and the other one for driving the self sensing actuator,
- low pass filters, which serve as anti-aliasing filters and are used to interface the analogic part to the digital one,
- a real time microprocessor based digital system, which consists of a dSpace DS1103.

A schematic of the experimental set up is shown in Figure 5.1.

The cantilevered beam is made of aluminum, whose parameters are listed in Table 5.1, and it is covered with a thin layer of electrical insulating paint. In fact, in order to make the measuring unit working, it is crucial to guarantee that the negative electrode of the self sensing piezoelectric actuator is insulated from the negative electrodes of the disturbance actuator and of the piezoelectric sensor. Moreover, a clamping mechanism is used to clamp the beam on one side.

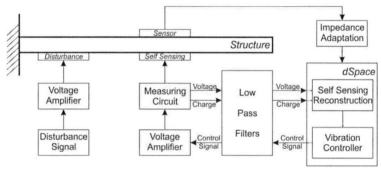

*Figure 5.1 – Schematic of the experimental set up for the cantilevered beam.*

| Parameter | Value |
|---|---|
| Young's modulus | 69 GPa |
| Length | 0.6 m |
| Width | 50 mm |
| Thickness | 3 mm |
| Poisson's ratio | 0.3 |

*Table 5.1 – Parameters of the cantilevered beam.*

| Parameter Symbol | Value |
|---|---|
| $\varepsilon_{33}^{\sigma}/\varepsilon_0$ | 2400 |
| $d_{31}$ | $-210 \cdot 10^{-12}$ C / N |
| $s_{11}^E$ | $15 \cdot 10^{-12}$ m$^2$ / N |
| $L$ | 50 mm |
| $b$ | 30 mm |
| $h$ | 0.2 mm |
| $\upsilon$ | 0.34 |

*Table 5.2 – Parameters of the piezoelectric plates (PIC151 from PI Ceramic).*

Three piezoelectric plates are bonded on the structure, as shown in Figure 5.1. Such actuators are provided by PI Ceramic and are made of the PIC 151 ceramic type. Their properties and parameters are listed in Table 5.2. Each piezoelectric plate is bonded by using an epossidic glue, which glues the negative electrode side to the structure.

A thin layer of copper foil is positioned between the piezoelectric transducers and the structure in order to access the negative electrodes. This layer does not cover the whole surface of the electrode but only a small part of it. Wires are then soldered on the copper foil for the negative electrodes, and on the positive electrodes by using a low temperature soldering procedure, to avoid damaging the thin layer of electrode deposited on the ceramic. A picture of the piezoelectric cantilevered beam is shown in Figure 5.2.

The measuring circuit used for these experiments is an Active Sawyer Tower circuit, which has been tuned according to the piezoelectric capacitance (measured at 1 kHz with a multimeter) and to the necessary amount of current and whose parameters are shown in Table 5.3.

The voltage amplifiers are bipolar and able to drive piezoelectric actuators in the range ±150 V. In fact, even though such actuators could be driven up to ±200 V, it is convenient to limit the maximum applicable voltage to avoid risks of crack in the very thin electrode layers due to high structural deformation. Furthermore, a voltage amplifier is used for driving the disturbance actuator (a function generator provides the reference signal), while the second one drives the self sensing actuator.

The digital system used for these experiments consists of a dSpace DS1103 module, which can be interfaced to Matlab/Simulink and runs both the self sensing reconstruction algorithm and the vibration control algorithm. Moreover, as for all digital implementations, low pass filters are necessary for antialiasing purposes. Such filters are realized with an active topology and have a cut off frequency at 5 kHz as the sampling frequency is 10 kHz. Finally, a Rhode&Schwarz audio analyzer has been used for measuring the frequency responses of the piezoelectric cantilevered beam.

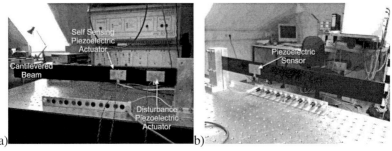

*Figure 5.2 – Piezoelectric cantilevered beam. a) front side; b) back side.*

| Parameter | Value |
|---|---|
| Measuring Capacitance $C_M$ | 2.2 µF |
| Measuring Resistance $R_M$ | 10 MΩ |
| Piezoelectric Capacitance $C_P$ | 96 nF |

*Table 5.3 – Parameters of the Active Sawyer Tower.*

## 5.1.2. Structural frequency response

In order to test and compare the linear and the hysteretic self sensing reconstruction, an analysis of the structure has been conducted. The structural frequency response of the cantilevered beam, obtained by using the self sensing actuator as a normal actuator, and by measuring the deformation of the structure by means of the coupled piezoelectric sensor, has been measured. Figures 5.3 and 5.4 show the measured frequency response driving the actuator respectively at low voltage (± 15 V) and at high voltage (± 150 V). Both frequency responses are obtained by providing a sweep driving voltage from 1 Hz up to 500 Hz and measuring the relative sensor voltage.

As one can see, the actuator/sensor frequency response does not change significantly over the amplitude of the driving voltage. As expected, a small contribution to the phase behaviour is present and increases over the frequencies that is due to the delays of the electronic measurement circuits.

The cantilevered beam presents 5 resonance frequencies related to bending modes in the analyzed frequency range, who are listed in Table 5.4 with the relative antiresonant frequencies.

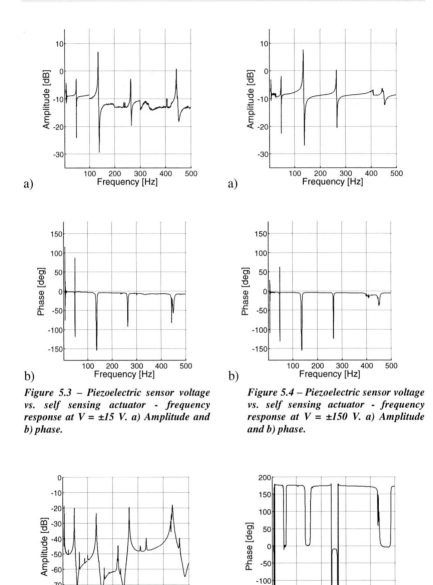

Figure 5.3 – Piezoelectric sensor voltage vs. self sensing actuator - frequency response at V = ±15 V. a) Amplitude and b) phase.

Figure 5.4 – Piezoelectric sensor voltage vs. self sensing actuator - frequency response at V = ±150 V. a) Amplitude and b) phase.

Figure 5.5 – Piezoelectric sensor voltage vs. disturbance actuator - frequency response. a) Amplitude and b) phase.

| Mode | Resonance Frequency [Hz] | Antiresonance Frequency [Hz] |
|:---:|:---:|:---:|
| 1 | 8.1 | 9.4 |
| 2 | 48.4 | 48.75 |
| 3 | 134.7 | 138.5 |
| 4 | 264.8 | 267.1 |
| 5 | 446.5 | 454.2 |

*Table 5.4 – Cantilevered beam resonance frequencies.*

The goal of this work is damping one resonance frequency by using a vibration controller with sensory information obtained by linear and hysteretic reconstruction. The third resonant frequency has been chosen as the frequency to be damped. In particular, the self sensing piezoelectric plate seems to have significant authority on the $3^{rd}$ mode due to its position on the beam.

It is important to remark that the sensor voltage is proportional to its deformation, which is the same of the self sensing actuator since they are bonded at the same location on the beam, one per each side. The self sensing reconstruction aims to reconstruct the deformation of the self sensing piezoelectric actuator. Thus a similar frequency response is expected when measuring the reconstructed deformation over voltage.

The frequency response shown in Figure 5.5 has been obtained by driving the disturbance actuator with a sweep driving voltage in the range ±150 V from 1 Hz up to 500 Hz and by measuring the structural deformation with the piezoelectric sensor.

## 5.1.3. Self sensing linear reconstruction

The value of the electrical clamped capacitance of the self sensing piezoelectric transducer must be known in order to perform a linear reconstruction with digital implementation. Consequently, as discussed in section 3.3.2, white noise has been applied to the self sensing piezoelectric actuator and, by means of a LMS algorithm, the electrical clamped capacitance value has been identified to be equal to 72 nF. This value has then been used for tuning the linear reconstruction algorithm on the digital system.

Once that the linear reconstruction algorithm is tuned on the electrical clamped capacitance of the self sensing actuation, it is ready to reconstruct the mechanical quantity of interest, i.e. the self sensing actuator deformation. Thus, again, two frequency responses are measured, which are shown in Figures 5.6 and 5.7, which are obtained by driving the self sensing actuator with a sweep voltage from 1 Hz up to 1 kHz respectively at low (±15 V) and high voltages (±150 V) and by reconstructing the piezoelectric deformation with the linear reconstruction algorithm.

The first consideration that can be made is that there is a strong change in the measured frequency responses at low and high driving voltages. In fact, while at low voltages the dynamic information of the structure can still be noticed, the frequency response amplitude and phase are practically flat at high voltages.

Focusing on the linear reconstruction at low driving voltages, the reconstructed phase diagram consists of a decreasing component over the frequencies added to a constant phase delay. While the first part is due to the phase delay introduced by the digital system and by the electronics, the second component of the phase delay is due to a mismatch of the linear electrical capacitance model and the real one, which is hysteretic. In fact, this introduces a constant phase delay of -10°, which decreases by increasing the amplitude of the driving voltage, thus by increasing the hysteretic loop width.

The phase shift introduced by the linear reconstruction is due to the uncompensated part of the electrical charge, which affects also the amplitude diagram and the amplitude of the phase shift corresponding to the resonances. In fact, comparing the frequency response in Figure 5.6 to the one in Figure 5.3, one can notice that the resonance peaks in the amplitude diagram, as well as the amplitude of the phase delay at the resonances, are smaller in the linear reconstruction than in the actuator / sensor case. By increasing the driving voltage, the uncompensated part of the electrical charge increases and becomes abundant with respect to the mechanical charge which is proportional to the piezoelectric deformation.

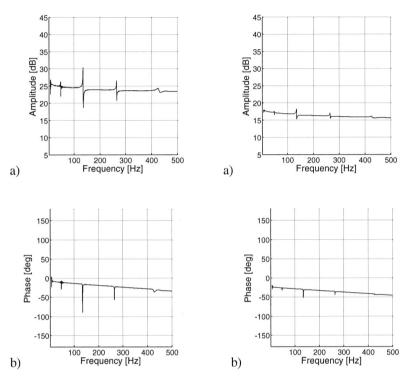

a)

a)

b)

b)

*Figure 5.6 – Linear reconstruction vs. self sensing actuator driving voltage - frequency response at V = ±15 V. a) Amplitude and b) phase.*

*Figure 5.7 – Linear reconstruction vs. self sensing actuator driving voltage - frequency response at V = ±150 V. a) Amplitude and b) phase.*

## 5.1.4. Self sensing hysteretic reconstruction

As discussed in section 3.5, a hysteretic reconstruction has been implemented by identifying the hysteretic characteristic of the electrical clamped capacitance by measuring it at the antiresonance frequency. In this case, the closest antiresonance frequency to the third resonance frequency is at 138.5 Hz (see Table 5.3). The identification voltage which is applied at the self sensing piezoelectric actuator is a sinusoidal voltage tuned at the antiresonance frequency and whose amplitude sweeps from low to high voltages in order to excite several hysteretic loops.

The mathematical model which has been chosen to model the hysteretic capacitance is the Modified Prandtl-Ishlinskii (see Appendix A). This model is a phenomenological model that can be identified with a quadratic program algorithm and by the measurements of voltage and charge. Moreover, the identification error of such model is typically around 1% ... 2%, which makes it very suitable for self sensing applications, where high modeling precision is needed. In this case, the identification error of the hysteretic model has given a 2.2% error.

The electrical clamped capacitance is rate-dependent, while the chosen mathematical model for this investigation models only non rate-dependent characteristics. Thus, the self sensing reconstruction is expected to give better performance in the neighbor frequencies of the antiresonance one. In the next section, instead, a rate dependent version of the Modified Prandtl-Ishlinskii is used for reconstructing the deformation of a piezoelectric actuator bonded on a plate structure.

As for the linear reconstruction, two frequency responses measured at low (±15 V) and at high voltages (±150 V), respectively, are shown in Figures 5.8 and 5.9.

Differently from the linear reconstruction, in this case the reconstructed dynamics gets closer to the frequency response of the actuator / sensor system shown in Figures 5.3 and 5.4 which has to be considered as the term of comparison for the reconstruction algorithms. In fact, the amplitude of the resonance peak is larger at the chosen resonance frequency, and the phase reconstruction goes down to -150°, as expected (see Figures 5.3 and 5.4). Larger amplitudes in amplitude peaks and phase are reconstructed also at other resonance frequencies and not only to the one associated to the antiresonance chosen for the identification of electrical clamped capacitance of the self sensing piezoelectric actuator.

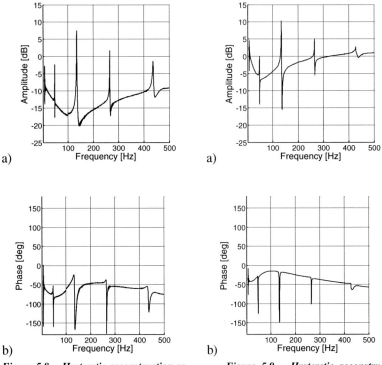

**Figure 5.8 – Hysteretic reconstruction vs. self sensing actuator driving voltage - frequency response at V = ±15 V. a) Amplitude and b) phase.**

**Figure 5.9 – Hysteretic reconstruction vs. self sensing actuator driving voltage - frequency response at V = ±150 V. a) Amplitude and b) phase.**

Moreover, the hysteretic reconstruction retrieves better information than the linear reconstruction both at low and at high driving voltages, although it is based on a hysteretic model which is affected by the systematic error of the uncompensated mechanical charge present at the antiresonance frequency chosen for the identification.

Furthermore, one can notice that the amplitude and phase diagrams are not flat, in fact a small curvature is noticeable in both diagrams. This is not only due to the biased identification of the electrical clamped capacitance, but also to the usage of a non rate dependent hysteretic model, which does not account for the change of the electrical clamped capacitance over the frequency due to creep effect.

## 5.1.5. Open loop response comparison

In this section a comparison of the open loop responses measured with the reference piezoelectric sensor and by using the linear and hysteretic reconstructions is briefly discussed.

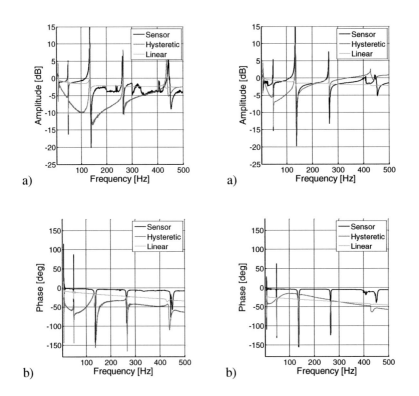

*Figure 5.10 – Open loop frequency response at V = ±15 V. a) Amplitude and b) phase.*

*Figure 5.11 – Open loop frequency response at V = ±150 V. a) Amplitude and b) phase.*

In Figures 5.10 and 5.11 the frequency responses, at low and at high driving voltages, shown in Figures 5.3, 5.4 and 5.6-5.9 are shown together in order to highlight the differences of the reconstruction methods.

As one can see, the linear reconstruction retrieves an open loop frequency response whose resonance peaks are smaller compared to the ones obtained with the reference sensor or by using the hysteretic reconstruction.

At high driving voltages this is even more evident, due to the static contribution given by the width of the uncompensated hysteretic electrical capacitance. This contribution is evident also in the phase reconstruction, where the error becomes relevant when compared to the reference sensor phase. Furthermore, while, on one hand, the hysteretic reconstruction is not affected by the level of the driving voltage, it is able to provide a closer match with the reference sensor only in the neighbor frequencies of the anti resonant frequency used for the hysteretic identification. This must not be considered a strong limitation, since vibration controllers need a higher accuracy in the frequency ranges of the resonance frequencies to be dampened. Thus, it is important to obtain a higher reconstruction accuracy at the resonance and antiresonance frequencies which is where, indeed, the hysteretic reconstruction succeeds.

## 5.1.6. Vibration control with self sensing reconstruction

In order to compare the linear and the hysteretic self sensing reconstructions in closed loop, a resonant controller has been implemented. The reconstructed signals are then fed back to the controller and used for damping the vibration induced by the disturbance actuator (see Figure 5.1). At the aim of testing the two reconstruction algorithms at low and high driving voltages, which are commanded by the controller, a low and a high disturbance is induced in the structure by driving the disturbance actuator respectively at ±40 V and ±150 V while sweeping in a frequency range around the resonant frequency of interest. In this way the vibration controller would command higher driving voltages to the self sensing actuator when a high disturbance is commanded on the structure. Finally, the piezoelectric sensor is used to measure the structural vibration.

The performance of the same vibration controller is shown both in the case of the linear and hysteretic reconstruction at low and at high disturbance in Figures 5.12 and 5.13, respectively.

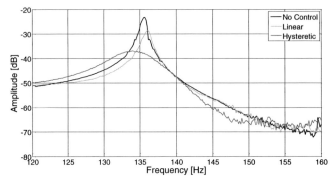

**Figure 5.12 – Sensor voltage vs. disturbance voltage – frequency response amplitude at low disturbance (±40 V) with resonant control.**

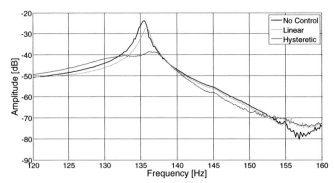

**Figure 5.13 – Sensor voltage vs. disturbance voltage – frequency response amplitude at high disturbance (±150 V) with resonant control.**

As one can see, in both cases the resonant controller performs better when receiving the sensory information from the hysteretic reconstruction rather than from the linear one. The hysteretic reconstruction, in fact, allows higher controller gains than the linear reconstruction. In this experiment, the resonant controller gain was set equal to 10 for the linear reconstruction and equal to 25 for the hysteretic one. The resonant controller, in fact, gets into instability with the linear reconstructor at gains larger than 10. Thus, the linear reconstruction sets a considerable gain limit beyond which the system becomes instable, and this is explainable by observing the small phase information that the linear reconstruction can provide (see Figures 5.6 and 5.7) differently from the hysteretic one (see Figures 5.8 and 5.9).

## 5.2. Partially clamped piezoelectric plate

Both linear and hysteretic self sensing techniques have been tested also on a partially clamped piezoelectric plate. As in the previous investigation, in this case a piezoelectric actuator is driven to induce a disturbance vibration into the structure while a self sensing actuator aims at reducing the structural vibration being driven by a vibration controller. Differently from the experiments on the cantilevered beam, where a piezoelectric sensor has been used, in this case a laser position sensor is used to measure the structural vibration.

Furthermore, two vibration controllers have been implemented in this experiment: a PPF controller and a resonant controller (see Appendix B).

### 5.2.1. Experimental set up description

The experimental set up consists of 6 parts:

- a partially clamped plate with two piezoelectric actuators bonded on it: the first one serves as a disturbance actuator while the second one as a self sensing actuator,
- a laser position sensor for measuring the out of plane deformation of the structure,
- a measuring circuit for measuring simultaneously the voltage and the charge of the self sensing actuator,
- two voltage amplifiers, one for driving the disturbance actuator and the other one for driving the self sensing actuator,
- low pass filters, which are used as anti-aliasing filters to interface

the analogic part to the digital one,

- a real time microprocessor based digital system, which consists of a dSpace DS1103.

A schematic of the experiment is shown in Figure 5.14.

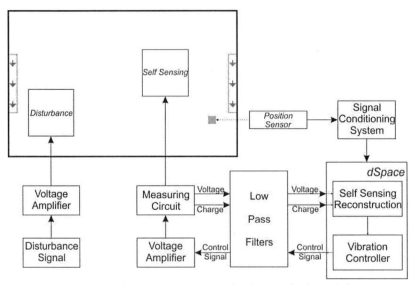

*Figure 5.14 – Schematic of the experimental set up for the partially clamped plate.*

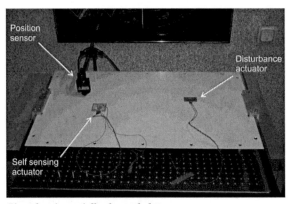

*Figure 5.15 – Piezoelectric partially clamped plate.*

The plate structure is made of aluminum, whose parameters are listed in

Table 5.5, and it is covered with a thin layer of electrical insulating paint to guarantee electrical insulation between the disturbance actuator and the self sensing one.

The plate structure is partially clamped on two sides, by letting the other two ones unconstrained, as shown in Figures 5.14 and 5.15. Two piezoelectric actuators have been bonded on the structure, which are made of different piezoelectric ceramics, and whose parameters are listed in Tables 5.6 and 5.7 for the self sensing and disturbance actuators, respectively.

| Parameter | Value |
|-----------|-------|
| Young's modulus | 69 GPa |
| Length | 0.8 m |
| Width | 0.5 m |
| Thickness | 3 mm |
| Poisson's ration | 0.3 |

*Table 5.5 – Parameters of the plate structure.*

The self sensing piezoelectric actuator, a Sonox P53 ceramic produced by CeramTec, is much thinner than the disturbance piezoelectric actuator, a PIC151 ceramic produced by PI Ceramic, for the reasons discussed in section 2.2. As in the previous work on the cantilevered beam, also in this case a thin layer of copper foil is positioned between the piezoelectric transducer and the structure in order to access the negative electrodes.

A laser position sensor is used for measuring the vertical displacement of the structure in one point, which has been chosen randomly and in a way that it is not a node for the vibration modes to be considered. This sensor is then connected to its signal conditioning system who provides the position signal to the digital system.

Also in this case, an Active Sawyer Tower cicuit has been tuned on the value of the capacitance of the piezoelectric self sensing actuator for measuring its voltage and charge, and the main parameters are listed in Table 5.8.

The voltage amplifier used for driving the self sensing actuator works in a range of ±150 V, while the voltage amplifier used for driving the disturbance actuator can provide voltages within a range of 0…800 V. The remaining electronics and the digital system used are the same as the ones shown in section 5.1.1.

| Parameter Symbol | Value |
|---|---|
| $\varepsilon_{33}^{\sigma}/\varepsilon_0$ | 3800 |
| $d_{31}$ | $-275\cdot10^{-12}$ C / N |
| $s_{11}^{E}$ | $15{,}8\cdot10^{-12}$ m$^2$ / N |
| $L$ | 60 mm |
| $b$ | 60 mm |
| $h$ | 0.27 mm |
| $\upsilon$ | 0.34 |

*Table 5.6 – Parameters of the self sensing piezoelectric actuator (Sonox P53 from CeramTec).*

| Parameter Symbol | Value |
|---|---|
| $\varepsilon_{33}^{\sigma}/\varepsilon_0$ | 2400 |
| $d_{31}$ | $-210\cdot10^{-12}$ C / N |
| $s_{11}^{E}$ | $15\cdot10^{-12}$ m$^2$ / N |
| $L$ | 50 mm |
| $b$ | 30 mm |
| $h$ | 1 mm |
| $\upsilon$ | 0.34 |

*Table 5.7 – Parameters of the disturbance piezoelectric actuator (PIC151 from PI Ceramic).*

| Parameter | Value |
|---|---|
| Measuring Capacitance $C_M$ | 2.2 µF |
| Measuring Resistance $R_M$ | 10 MΩ |
| Piezoelectric Capacitance $C_P$ | 192 nF |

*Table 5.8 – Parameters of the Active Sawyer Tower.*

## 5.2.2. Structural frequency response

In order to retrieve some information about the structure, its frequency response has been measured by driving the disturbance actuator with a sweep driving voltage from 1 Hz up to 500 Hz at an amplitude of 800 V peak to peak and offset at 400 V and measuring the vertical displacement of the structure in one point with the laser position sensor (Figure 5.16).

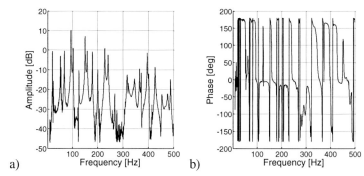

***Figure 5.16 – Laser sensor voltage vs. disturbance actuator - frequency response. a) Amplitude and b) phase.***

As one can observe, many resonances are present in the frequency window between 1 Hz and 500 Hz. In order to test the quality of the self sensing reconstruction algorithms, two resonance frequencies are chosen to be damped, one at the low region and the other one at the high region of the frequency range, and they are respectively at 94.2 Hz and at 396.3 Hz. They have been chosen for their high peaks. In fact, the resonance frequency at 94.2 Hz has the highest peak among the others in the range 1 ... 250 Hz, as well as the second chosen resonance frequency at 396.3 Hz has the highest peak in the range 250 ... 500 Hz.

Differently from the previous work on the cantilevered beam, a piezoelectric sensor coupled to the self sensing actuator is not present since no suitable piezoelectric samples were available, and consequently one does not have a reference frequency response like the one shown in Figures 5.3 and 5.4 to be compared to the self sensing reconstructed frequency responses.

## 5.2.3. Self sensing linear reconstruction

The linear identification of the electrical clamped capacitance by noise injection in the self sensing actuator, as discussed in section 3.3.2, has led to a value equal to 159.7 nF. This value has been obtained by using both the LMS and the RLS algorithms. The linear reconstruction algorithm implemented on the digital system has then been tuned on this value.

Two frequency responses have been measured by driving the self sensing actuator with a sweep voltage from 1 Hz up to 500 Hz and using the linear reconstructor unit. The first frequency response is obtained at low driving voltages (±8 V) while the second one at high driving voltages (±100 V), and their amplitude and phase graphs are respectively shown in Figures 5.17 and 5.18. In this work lower driving voltages than in the previous example of the cantilevered beam have been used in order to protect the piezoelectric ceramics from damages due to the high vibrational excitation they are exposed to.

As one can see, the amplitude graphs change significantly (note that the amplitudes of the axis scale of Figures 5.15 and 5.16 are different). While at low voltages the structural dynamics is still evident, at high voltages the amplitude graph is almost flat and it is shifted of ca. +10 dB. Thus the dynamic behavior of the structure does not appear. This is due to the uncompensated electrical charge which becomes consistent at high voltages. A strong reduction of the phase dynamic content can be observed also in the phase diagrams. Let us then recall equation (3.53):

$$G_v^{\dot{q}_m}(s) = \frac{\mathbf{C}^q Adj(s\mathbf{I}-\mathbf{A})\mathbf{B}_{rv}}{|s\mathbf{I}-\mathbf{A}|} + \frac{\tilde{C}_p|s\mathbf{I}-\mathbf{A}|}{|s\mathbf{I}-\mathbf{A}|} = G_v^{\dot{q}_m}(s) + \tilde{C}_p. \qquad (4.1)$$

The strong change in the frequency response at high voltages respect to the one at low voltages is due in fact to the capacitance identification error $\tilde{C}_p$ which increases at high driving voltages and consequently moves the zeros closer to the poles towards a condition of zeros/poles cancellation. In such condition, the reconstructed frequency response tends to be flat in amplitude and phase as a static system.

Moreover, the reconstructed frequency response at high driving voltages shows a larger phase delay (around -20°) which is due to the fact that the hysteretic curve of the electrical clamped capacitance becomes wider. Consequently, a linear model results in a constant phase delay which is proportional to the width of the hysteretic curve (see Figures 5.17(b) and 5.18(b)).

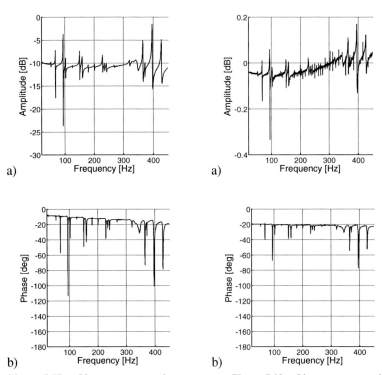

a)

a)

b)

b)

*Figure 5.17 – Linear reconstruction vs. self sensing actuator driving voltage - frequency response at V = ±8 V. a) Amplitude and b) phase.*

*Figure 5.18 – Linear reconstruction vs. self sensing actuator driving voltage - frequency response at V = ±100 V. a) Amplitude and b) phase.*

## 5.2.4. Self sensing hysteretic reconstruction

In order to perform hysteretic reconstruction, two hysteretic models have been identified, one per each resonant frequency of interest. The identification of the two hysteretic models has been performed by measuring the piezoelectric capacitance at the closest antiresonances, which are at 95.3 Hz and at 399.7 Hz.

Two Modified Prandtl-Ishlinskii models have been used for modeling the piezoelectric capacitances at the two antiresonances (see Appendix A). Differently from the previous experiment on the cantilevered beam, in this case the hysteretic models are rate dependent, since also creep effect has been considered (see section 2.1.3). In order to perform the identification, the self sensing actuator has been driven with a sinusoidal voltage whose amplitude increases from ±5 V up to ±100 V. The two hysteretic models have been then identified with an error of 1.2% and 1.8% (see equation A.20), respectively. Figures 5.19 and 5.20 show the identified hysteretic characteristics.

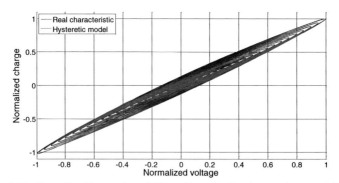

*Figure 5.19 – Normalized hysteretic modeling of the piezoelectric capacitance at 95.3 Hz.*

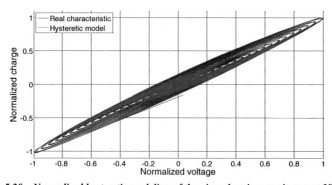

*Figure 5.20 – Normalized hysteretic modeling of the piezoelectric capacitance at 399.7 Hz.*

Four frequency responses have been measured by using alternatively the two hysteretic models at low (±8 V) and at high (± 100 V) driving voltages, and they are shown in Figures 5.21 ... 5.24, respectively.

This experiment shows also that increasing the driving voltage does not affect considerably the reconstructed frequency response, which is one of the first aims to pursue. As in the case of the cantilevered beam, the hysteretic reconstruction provides larger resolution in terms of frequency response amplitudes and phases than the linear reconstruction, even though the identification data lead to a biased identification of the hysteretic capacitance. In fact, amplitude peaks are larger and phase delays go down to -180° in correspondence with the resonant frequencies as expected from theory.

Moreover, one can notice that the reconstructed frequency responses are more precise near the antiresonances used for identification. Let us focus on the frequency responses shown in Figures 5.21 and 5.22. As one can see the amplitude of the peaks at high frequencies (near the antiresonance at 399.7 Hz) are smaller than expected, as they cover only a few dBs range, as well as the corresponding phase delays, while peaks and phase delays are larger near the antiresonance at 95.3 Hz, which is used for the identification of the hysteretic model used for reconstruction. Similar consideration can ben done in the case of the frequency responses of Figures 5.23 and 5.24, where amplitude and phase delays are larger at the frequencies near the chosen antiresonance at 399.7 Hz, while at low frequencies the amplitudes of the resonance frequencies are smaller than expected (small reconstructed dynamics) and the phase delays are shifted up to a maximum of +150°. The phase delay gets then close to zero in proximity of the antiresonance at 399.7 Hz.

These observations lead to consider that the creep effect which has been modeled in this example is not sufficient to describe the frequency dependency of the piezoelectric capacitance, and consequently it is convenient to use the hysteretic reconstruction for vibration control only in a frequency range close to the antiresonance chosen for identification of the piezoelectric capacitance.

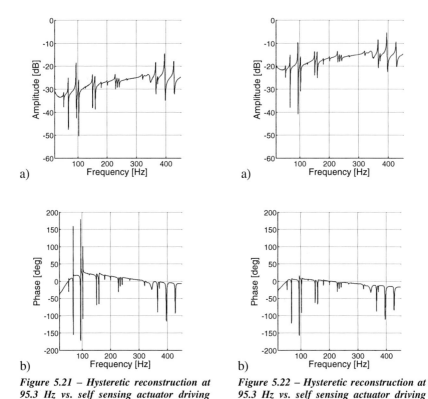

a)

b)

**Figure 5.21 – Hysteretic reconstruction at 95.3 Hz vs. self sensing actuator driving voltage - frequency response at V = ±8 V. a) Amplitude and b) phase.**

**Figure 5.22 – Hysteretic reconstruction at 95.3 Hz vs. self sensing actuator driving voltage - frequency response at V = ±100 V. a) Amplitude and b) phase.**

### 5.2.5. Vibration control with self sensing reconstruction

Two controllers have been implemented in order to prove the quality of the linear and hysteretic reconstruction algorithms: a PPF controller and a resonant controller. As shown in Figure 5.14, the disturbance actuator induces a disturbance vibration into the structure, while the self sensing actuator is driven by the controller in order to reduce the structural vibration. Finally the laser position sensor is used to measure the amount of structural transversal vibration. The two controllers have been tuned in order to damp the two resonance frequencies which are at 94.2 Hz and 396.3 Hz. The performance of the two reconstruction algorithms is then tested both at low and at high disturbance vibrations, which are obtained by driving respectively the disturbance actuator at 100 V and 800 V peak to peak amplitude with an offset of 400 V.

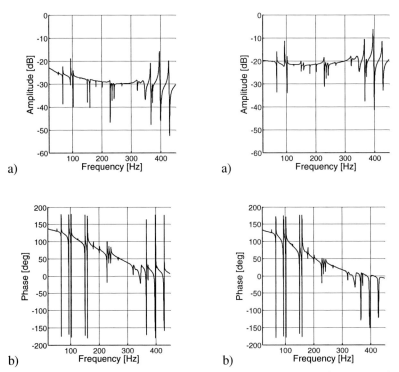

a)

b)

*Figure 5.23 – Hysteretic reconstruction at 399.7 Hz vs. self sensing actuator driving voltage - frequency response at V = ±8 V. a) Amplitude and b) phase.*

*Figure 5.24 – Hysteretic reconstruction at 399.7 Hz vs. self sensing actuator driving voltage - frequency response at V = ±100 V. a) Amplitude and b) phase.*

Concerning the hysteretic reconstruction, the hysteretic model identified at 95.3 Hz is used for reconstruction in the frequency range of the resonance at 94.2 Hz while the hysteretic model identified at 399.7 Hz is used for reconstruction in the frequency range of the resonance at 396.3 Hz. Figures 5.25 … 5.32 show the obtained measurements.

151

These figures show clearly how the hysteretic reconstruction allows better control performance than the linear one, not only at high driving voltages, but also at low ones. In fact, even at low driving voltages the electrical clamped capacitance still exhibits a hysteretic loop which leads to a constant phase delay which deteriorates the quality of the reconstruction. This effect is noticeable in Figures 5.17 and 5.18, where a constant phase shift is clearly viewable.

The experimental results shown in this chapter are coherent with the theoretical investigations provided in Chapter 3.

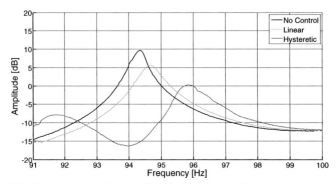

*Figure 5.25 – Laser sensor voltage vs. disturbance voltage – frequency response amplitude at low disturbance (100 V amplitude) with PPF control tuned at 94.2 Hz.*

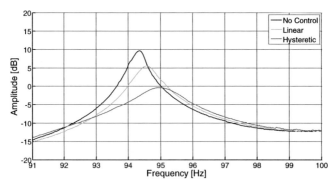

*Figure 5.26 – Laser sensor voltage vs. disturbance voltage – frequency response amplitude at low disturbance (100 V amplitude) with resonant control tuned at 94.2 Hz.*

152

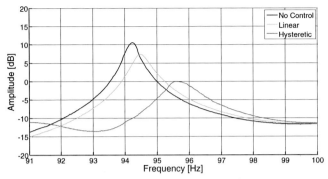

*Figure 5.27 – Laser sensor voltage vs. disturbance voltage – frequency response amplitude at high disturbance (800 V amplitude) with PPF control tuned at 94.2 Hz.*

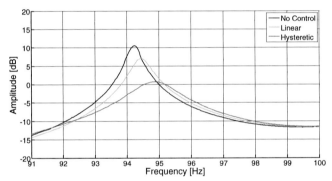

*Figure 5.28 – Laser sensor voltage vs. disturbance voltage – frequency response amplitude at high disturbance (800 V amplitude) with resonant control tuned at 94.2 Hz.*

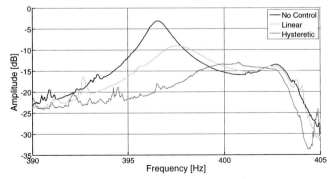

*Figure 5.29 – Laser sensor voltage vs. disturbance voltage – frequency response amplitude at low disturbance (100 V amplitude) with PPF control tuned at 396.3 Hz.*

153

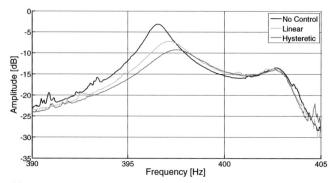

*Figure 5.30 – Laser sensor voltage vs. disturbance voltage – frequency response amplitude at low disturbance (100 V amplitude) with resonant control tuned at 396.3 Hz.*

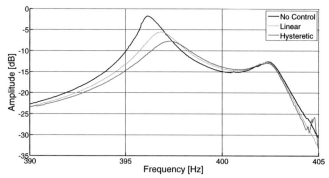

*Figure 5.31 – Laser sensor voltage vs. disturbance voltage – frequency response amplitude at high disturbance (800 V amplitude) with PPF control tuned at 396.3 Hz.*

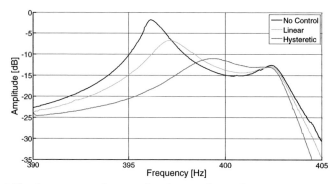

*Figure 5.32 – Laser sensor voltage vs. disturbance voltage – frequency response amplitude at high disturbance (800 V amplitude) with resonant control tuned at 396.3 Hz.*

## 5.3. Comparison between linear and hysteretic reconstructions

As experimentally shown, the hysteretic reconstruction allows better vibration control performance at all driving voltages. Nevertheless it is opportune to remark that both linear and hysteretic reconstructions present advantages and drawbacks.

The linear reconstruction is easily implementable. In fact, as discussed in section 3.6, it can be implemented analogically, digitally and in a hybrid way, and thus it is highly versatile. Moreover, the possibility of performing adaptive identification of the electrical clamped capacitance simplifies considerably the effort in identifying the piezoelectric electrical clamped capacitance, which is not identifiable with direct measurements. Furthermore, adaptive identification is able to tune the bridge circuit during operation, which is important when the piezoelectric parameters change over the time, due to temperature variation or aging.

Nevertheless, adopting a linear reconstruction means being able to drive the self sensing piezoelectric actuator only at very low voltages and not being able to use the whole actuating capability of the device. Thus, it is possible that in some applications the number of needed self sensing actuators to achieve certain damping performance could be so high that a coupled piezoelectric actuator/sensor system would result to be more performant and cheap.

Hysteretic reconstruction solves this issue, since it is able to perform good reconstruction at low and at high driving voltages as the nonlinearities of the piezoelectric clamped capacitance are taken into consideration, although it is based on a biased identification due to the uncompensated mechanically induced charge present at antiresonance. In case it is important to use a self sensing actuator at its full actuating capability, such reconstruction assumes a crucial importance.

However, a hysteretic reconstruction algorithm requires a digital system to compute a hysteretic mathematical model. The analogic implementation of a hysteretic reconstruction would require a considerable effort and it has not been developed at this time. Moreover, the hysteretic reconstruction is based on the identification at the antiresonance frequencies, which have to be located with high precision. This task requires external structural investigations which can be burdensome, and it represents the weakest aspect of this technique.

# 6.   Conclusion

This work addresses the problem of using self sensing piezoelectric actuators at full driving voltages for vibration control purposes.

In order to get a full understanding of self sensing techniques, the main notions about piezoelectricity and vibrating structures have been given in Chapter 2. After presenting piezoelectric features and some main aspects of vibrating structures, a mathematical model for piezoelectric vibrating structures has been presented. Such working framework provides a necessary background for self sensing techniques.

In Chapter 3 the main self sensing techniques present in the scientific literature are discussed. They are all based on modelling the electrical clamped capacitance as a linear characteristic. This modelling choice provides good results only at low driving voltages, and the need of a self sensing technique capable of performing a reliable reconstruction at high driving voltages becomes consistent.

For this reason, a new self sensing technique based on a hysteretic model has been presented. Such technique compensates for the nonlinearities of the self sensing actuator and consequently sets no theoretical limits to the amplitude of the driving voltage.

In Chapter 4 the necessary electronics for performing self sensing reconstructions has been presented and discussed. A new active circuitry for the measurement of the electrical charge stored on a piezoelectric transducer and of the electric field strength applied during operation has been presented, which is more versatile and performant than the typical passive circuit well known in literature as Sawyer Tower circuit. Moreover, a small modification of this circuit has been introduced, which allows to measure also the temperature of a piezoelectric transducer.

In Chapter 5 two vibrating structures are used as experimental benches for testing the new self sensing technique and compare it to a typical linear technique. Experimental results have demonstrated how the hysteretic reconstruction not only has good performance at all driving voltages, but it even allows better control performance at low driving voltages, where the linear self sensing techniques are expected to perform well.

Nevertheless, self sensing techniques present always advantages and drawbacks and, consequently, it is not possible to assert that one technique is always better than another one. In fact, a weak aspect of the hysteretic reconstruction is the identification of the hysteretic piezoelectric capacitance, which is identified by measuring voltage and charge on the self sensing actuator by supplying a harmonic voltage signal tuned at one of the antiresonances of the piezoelectric vibrating structure. The identification procedure presents two main limitations:

- structural antiresonances have to be identified a priori,
- the identification algorithm of the hysteretic capacitance runs offline.

Thus, there is no possibility at the actual state of the art to perform an adaptive identification of the hysteretic characteristic, which would be useful for many practical applications, where temperature or aging could change the piezoelectric parameters over the time in a significant way. Therefore deeper investigations must be conducted in order to provide an online identification of the hysteretic electrical clamped capacitance, opening a new scenario for further research.

Besides, the influence of temperature on the self sensing reconstruction performance is not always neglectable. In many applications, like in the automotive sector as well as in many industrial environments, temperature plays a key role. Although this aspect has not been developed during this work, a first approach has been provided by presenting the Active Sawyer Tower with temperature measurement presented in Section 4.4. Such circuit, in fact, has been developed with the aim of being used for temperature affected self sensing reconstructions, which make use of vectorial hysteretic mathematical models. Thus, another important scenario for further research should consist in developing new self sensing algorithms capable of facing temperature changes of the self sensing piezoelectric transducer.

# Bibliography

[AH94]      Anderson, E. H. and Hagood, N. W. (**1994**). *Simultaneous piezoelectric sensing/actuation: analysis and application to controlled structures*. Journal of Sound and Vibration, Volume 174, Number 5, 617-639.

[And86]      Anderson, E. H. (**1986**). *Piezoceramic Induced Strain Actuation of One- and Two-Dimensional Structures*. Master of Science in Aeronautics and Astronautics, Massachussets Institute of Technology.

[CA90]      Crawley, E. F. and Anderson, E. H. (**1990**). *Detailed Models of Piezoceramic Actuation of Beams*. Journal of Intelligent Material Systems and Structures. Volume 1, Number 4, 4-25.

[CC80]      Curie, J. and Curie, P (**1880**). *Contractions et dilatations produites par des tensions dans les cristaux hémièdres à faces inclinées.* C R Acad Sci Gen., 93:1137–1140.

[CC94]      Cole, D. G. and Clark, R. L. (**1994**). *Adaptive Compensation of Piezoelectric Sensoriactuators*. Journal of the Acoustical Society of America, Volume 95, Issue 5, 2989.

[CKK68]      Crandall, S. H., Karnopp, D. C., Kurtz, E. F., Pridmore-Brown, D. C. (**1968**). *Dynamics of Mechanical and Electromechanical Systems*. Robert E. Krieger Publishing Co., Malabar, Florida, 291-326.

[CL87]      Crawley, E. F., de Luis, J. (**1987**). *Use of Piezoelectric Actuators as Elements of Intelligent Structures*. AIAA Journal, Volume 25, Number 10, 1373-1385.

[DIG92]      Dosch, J. J., Inman, D. J., Garcia, E. (**1992**). *A Self-Sensing Piezoelectric Actuator for Collocated Control*. Journal of Intelligent Material Systems and Structures, Volume 3, 166-185.

[FC90]      Fanson, J. L. and Caughey, T. K. (**1990**). *Positive Position Feedback Control for Large Space Structures*. AIAA Journal, Volume 28, Number 4, 717–724.

[FL08]      Fenik, S. and Ladislav, S. (**2008**). *Optimal PPF Controller for Multimodal Vibration Suppression Engineering Mechanics*. Engineering Mechanics, Volume 15, Number 3, 153-173.

[GMJ11]     Grasso, E., May, C., Janocha, H., Naso, D. (**2011**). *Reducing force harmonics from a pendulum actuator*. Proc. Smart Structures and Materials (V Eccomas Thematic Conf.), Saarbrücken, Germany, 6-8 July 2011.

[GMJ12]     Grasso, E., May, C., Janocha, H., Naso, D. (**2012**). *Generating periodic forces with the pendulum actuator*. Journal of Vibration and Control. Volume 18, Number 1, 3-16.

[GPJ10]     Grasso, E., Pesotski, D., Janocha, H. (**2010**). *Self-Sensing Piezoelectric Actuators for Vibration Control Purposes*. Actuator 2010, Proc. 12th International Conference on New Actuators, (Bremen 14-16 June 2010), 151-154, ISBN 978-3-933339-13-3.

[GTJ13]     Grasso, E., Totaro, N., Janocha, H., Naso, D. (**2013**). *Piezoelectric self sensing actuators for high voltage excitation*. Journal of Smart Materials and Structures. Volume 22, Number 6.

[Hay91]     Haykin, S. (**1991**). *Adaptive Filter Theory, Second Edition*. Englewood Cliffs, NJ, Prentice Hall, 1991.

[HCF90]     Hagood, N. W., Chung, W. H., Von Flotow, A. (**1990**). *Modelling of Piezoelectric Actuator Dynamics for Active Structural Control*. Journal of Intelligent Material Systems and Structures. Volume 1, Number 3, 327-354.

[HMJ01]     Harland, N. R., Mace, B. R., Jones, R. W. (**2001**). *Adaptive-passive control of vibration transmission in beams using electro/magnetorheological fluid filled inserts*. IEEE Transactions on Control Systems Technology, Volume 9, Issue 2, 209-220.

[HSJ10]      Hong, C., Shin, C., Jeong, W. (**2010**). *Active vibration control of clamped beams using PPF controllers with piezoceramic actuators*. Proceedings of the 20th International Congress on Acoustics, ICA 2010, Sydney, Australia.

[Jan04]      Janocha, H., Ed. (**2004**). *Actuators – Basics and Applications*. Berlin, Heidelberg, New York: Springer-Verlag. ISBN: 3-540-61564-4.

[Jan07]      Janocha, H., Ed. (**2007**). *Adaptronics and Smart Structures. Basics, Materials, Design and Applications. Second, Revised Edition*. Berlin, Heidelberg, New York: Springer-Verlag. ISBN: 978-3-540-71965-6.

[JG97]       Jones, L. D., Garcia, E. (**1997**). *Novel approach to self-sensing actuation*. Smart Structures and Materials, Proc. SPIE 3041.

[JK06]       Janocha, H. and Kuhnen, K. (**2006**). *Self-Sensing Effects in Solid-State Actuators*. Grimes, C.A., Dickey, E.C., Pishko, M.V. (Ed.): Encyclopedia of Sensors, American Scientific Publishers, Volume 9, 53-74.

[Jon75]      Jones, R. M. (**1975**). *Mechanics of Composite Materials*. Scripta, Washington D.C.

[JQW11]      Ji, H. L., Qiu, J. H., Wu, Y. P., Cheng, J., Ichchou, M. N. (**2011**). *Novel approach of self-sensing actuation for Active Vibration Control*. Journal of Intelligent Material Systems and Structures, Volume 22, Number 5.

[KP89]       Krasnosel'skii, M. A. and Pokrovskii, A. V. (**1989**). *Systems with Hysteresis*. New York: Springer-Verlag. ISBN: 0-387-15543-0.

[KS10]       Kuiper, S. and Schitter, G. (**2010**). *Active damping of a piezoelectric tube scanner using self-sensing piezo actuation*. Mechatronics, Volume 20, Number 6, 656-665.

[Kuh01]      Kuhnen, K. (**2001**). *Inverse Steuerung piezoelektrischer Aktoren mit Hysterese-, Kriech- und Superpositionsoperatoren*. Shaker-Verlag. ISBN: 3-8265-9635-8.

[Kuh03]   Kuhnen, K. (**2003**). *Modeling, Identification and Compensation of Complex Hysteretic Nonlinearities. A modified Prandtl-Ishlinskii Approach.* European Journal of Control, Volume 9, Number 4, 407-418.

[Lee90]   Lee, C. K. (**1990**). *Theory of laminated piezoelectric plates for the design of distributed sensors/actuators. Part I: Governing equations and reciprocal relationships.* Journal Acoustic Society Am., Volume 87, Number 3, 1144-1158.

[Lei69]   Leissa, A. W. (**1969**). *Vibration of Plates.* NASA SP-160.

[LM90]   Lee, C. K., Moon, F. C. (**1990**). *Modal Sensors/Actuators.* Journal of Applied Mechanics, Transactions of the ASME, Volume 57, 434-441.

[MA10]   Mahmoodi, S. N. and Ahmadian, M. (**2010**). *Modified acceleration feedback for active vibration control of aerospace structures.* Smart Materials and Structures, Volume 19, 1-10.

[Mas81]   Mason, W. P. (**1981**). *Piezoelectricity, its history and applications.* Journal of Acoustic Society America, Volume 70, Number 6, 1561-1566.

[Mce99]   McEver, M. A. (**1999**). *Optimal Vibration Suppression Using On-line Pole/Zero Identification.* Master Thesis, Blacksburg: Virginia Tech.

[MJG09]   May, C., Janocha, H., Grasso, E., Naso, D. (**2009**). *A pendulum actuator and its force generation capabilities.* Proc. ASME 2009, IDETC/CIE 2009 (San Diego, USA, 30.08-02.09.2009).

[MTA07]   Miclea, C., Tanasoiu, C., Amarande, L., Miclea, C. F., Plavitu, C., Cioangher, M., Trupina, L., Miclea, C. T., David, C. (**2007**). *Effect of Temperature on the Main Piezoelectric Parameters of A Soft PZT Ceramic.* Romanian Journal of Information Science and Technology. Volume 10, Number 3, 243-250.

[New05]   Newnham, R. E. (**2005**). *Properties of materials. Anysotropy, symmetry, structure.* Oxford University Press. ISBN: 0-19-852075-1.

[PHS92]     Pan, J., Hansen, C. H., Snyder, S.D. (**1992**). *A Study of the Response of a Simply Supported Beam to Excitation by a Piezoelectric Actuator*. Journal of Intelligent Materials and Structures, Volume 3, Number 3.

[Pre02]     Preumont, A. (**2002**). *Vibration Control of Active Structures. An Introduction, 2^nd Edition*. Kluwer Academic Publishers, Dordrecht. ISBN 0-7923-4392-1.

[RC00]      Reza Moheimani, S. O., Clark, R. L. (**2000**). *Minimizing the truncation error in assumed modes models of structures. Revised Version*. American Control Conference. Volume 4, 2398-2402.

[RH03]      Reza Moheimani, S. O., Halim, D. (**2003**). *An optimization approach to optimal placement of collocated piezoelectric actuators and sensors on a thin plate*. Mechatronics, Volume 13, 27-47.

[RHF03]     Reza Moheimani, S. O., Halim, D., Fleming, A. J. (**2003**). *Spatial Control of Vibration – Theory and Experiments*. World Scientific Publishing. ISBN: 981-238-337-9.

[RPS99]     Reza Moheimani, S. O., Pota, H. R., Smith, M. (**1999**). *Resonant controllers for flexible structures*. Proceedings IEEE of the 38th International Conference on Decision and Control, Phoenix, Arizona.

[RVB06]     Reza Moheimani, S. O., Vautier, B. J. G., Bhikkaji, B. (**2006**). *Experimental Implementation of Extended Multivariable PPF Control on an Active Structure*. IEEE Transactions on Control Systems Technology, Volume 14, Number 3, 443-455.

[SHM04]     Simmers, G. E., Hodgkins, J. R., Mascarenas, D. D., Park, G., Sohn, H. (**2004**). *Improved Piezoelectric Self-sensing Actuation*. Journal of Intelligent Material Systems and Structures, Volume 15, 941-953.

[Smi05]     Smith, R. C. (**2005**). *Smart Material Systems – Model development*. Society for Industrial and Applied Mathematics.

[SSA01]     Song, G., Schmidt, S. P., Agrawal, B. N. (**2001**). *Experimental Robustness Study of Positive Position Feedback Control for Active*

*Vibration Suppression*. Journal Guidance, Volume 25, Number 1.

[ST30]     Sawyer, C. B. and Tower, C. H. (**1930**). *Rochelle Salt as a Dielectric*. Physical Review, Volume 35, 269–273.

[TS06]     Tjahyadi, F. H. and Sammut, K. (**2006**). *Multi-mode vibration control of a flexible cantilever beam using adaptive resonant control*. Smart Materials and Structures, Volume 15, 270–278.

[TW59]     Timoshenko, S. and Woinowsky-Kreiger, S. (**1959**). *Theory of Plates and Shells*. McGraw Hill. New York.

[Urg10]    Urgessa, G. S. (**2010**). *Vibration properties of beams using frequency-domain system identification methods*. Journal of Vibration and Control, Volume 17, Number 9, 1287-1294.

[VC95]     Vipperman, J. S. and Clark, R. L. (**1995**). *Hybrid analog and digital adaptive compensation of piezoelectric sensoriactuators*. AIAA-95-1098-CP, 2854-2859.

[VC96a]    Vipperman, J. S. and Clark, R. L. (**1996**). *Implementation of an Adaptive Piezoelectric Sensoriactuator*. AIAA Journal, Volume 34, Issue 10, 2102-2109.

[VC96b]    Vipperman, J. S. and Clark, R. L. (**1996**). *Complex adaptive compensation of nonlinear piezoelectric sensoriactuators*. American Institute of Aeronautics and Astronautics, AIAA Meeting Papers on Disc, 1996, 1-11.

[Voi10]    Voigt, W. (**1910**). *Lehrbuch der Kristallphysik*. B. G. Teubner, Leipzig, Berlin.

# A. A modified Prandtl-Ishlinskii model

Hysteresis is a non-linear phenomenon that is also called "rate-independent memory effect". This means that the present output value of a system with hysteresis depends on the present value of the input and on the history of the input, but not on its rate [Kuh03].

A possible way to model these types of nonlinearities is to use phenomenological models. They describe the input-output hysteretic curve without explicit reference to the physical phenomena that causes the nonlinear behaviour. The Modified Prandtl-Ishlinskii (MPI) model belongs to this category. In particular it is a modified version of the Prandtl-Ishlinskii (PI) model.

## A.1. Definition of the MPI model

The Prandtl-Ishlinskii (PI) model is composed by the weighted sum of $n+1$ elementary operators called *play operators*. Let $x$ be the input, $y_i$ the output and $r_{H_i}$ the threshold of the $i$-th play operator (see Figure A.1).

The output of $i$-th play operator for a generic time instant $t$ is:

$$ y_i(t) = H_i \left[ x, y_i, r_{H_i} \right](t) = \max \left\{ x(t) - r_{H_i}, \min \left\{ x(t) + r_{H_i}, y_i(t) \right\} \right\}. \quad \text{(A.1)} $$

The weighted sum of all PI operators, and consequently the output of the PI model, can be defined as:

$$ H[x](t) \triangleq \mathbf{w}_H^T \mathbf{H}_{\mathbf{r}_H} \left[ x, \mathbf{z}_{H_0} \right](t), \quad \text{(A.2)} $$

where $\mathbf{w}_H^T = \left( w_{H_0}, w_{H_1}, \ldots, w_{H_n} \right)$ is a vector of weights, $\mathbf{r}_H^T = \left( r_{H_0}, r_{H_1}, \ldots, r_{H_n} \right)$ is the vector of thresholds, with the property that $0 < r_{H_0} < r_{H_1} < \cdots < r_{H_n} < +\infty$, $\mathbf{z}_{H_0}^T = \left( z_{H_{00}}, z_{H_{01}}, \ldots, z_{H_{0n}} \right)$ is the vector of initial states while the vector or play operators is defined as

$$\mathbf{H}_{r_H}\left[ x, \mathbf{z}_{H_0} \right](t)^T = \left( H_{r_{H_0}}\left[ x, z_{H_{00}} \right](t), H_{r_{H_1}}\left[ x, z_{H_{01}} \right](t), \ldots, H_{r_{H_n}}\left[ x, z_{H_{0n}} \right](t) \right).$$

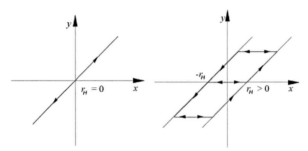

*Figure A.1 – x-y trajectory of the play operator.*

Assuming that the threshould values and the initial states are defined a priori, one can notice from equation (A.2) that the PI model is linear in the parameters, which in this case are the weights. The threshould values are usually distributed uniformly over the positive range of the input signal, while the initial states are typically set to zero.

Such model is suitable for identifying symmetric hysteresis, while it fails in identifying asymmetrical ones. For this reason, the PI model is extended by using a *superposition* operator, which is a memory-free operator.

Let $x$ be the input, $y_i$ the output and $r_{S_i}$ the threshould of the $i$-th superposition operator, the elementary superposition operator is the one-sided dead-zone operator $S_i$, and $y_i(t) = S_i\left( x(t), r_{S_i} \right)$. The elementary superposition operator is defined as:

$$S_i\left(x(t),r_{S_i}\right)=\begin{cases}\max\left\{x(t)-r_{S_i},0\right\} & r_{S_i}>0\\ x(t) & r_{S_i}=0\\ \min\left\{x(t)-r_{S_i},0\right\} & r_{S_i}<0\end{cases}.\qquad(A.3)$$

Figure A.2 shows the rate-independent output-input trajectory of the elementary superposition operator for different threshould values.

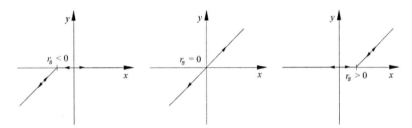

*Figure A.2 – x-y trajectory of the superposition operator.*

The superposition operator is then defined as the sum of $2l+1$ elementary superposition operators:

$$S[x](t)\triangleq \mathbf{w}_S^T\mathbf{S}_{\mathbf{r}_S}[x](t),\qquad(A.4)$$

where $\mathbf{w}_S^T=\left(w_{S_{-l}},\dots,w_{S_{-1}},w_{S_0},w_{S_1},\dots,w_{S_l}\right)$ is the vector of weights, $\mathbf{r}_S^T=\left(r_{S_{-l}},\dots,r_{S_{-1}},r_{S_0},r_{S_1},\dots,r_{S_l}\right)$ is the vector of threshould values and $\mathbf{S}_{\mathbf{r}_S}[x](t)^T=\left(S_{r_{S_{-l}}}[x](t),\dots,S_{r_{S_{-1}}}[x](t),S_{r_{S_0}}[x](t),S_{r_{S_{+1}}}[x](t),\dots,S_{r_{S_l}}[x](t)\right)$ is the vector of operators.

The MPI model is obtained by combining the PI model with the superposition operator, and it is defined as:

$$\Gamma[x](t)\triangleq S\left[H[x](t)\right](t)=\mathbf{w}_S^T\mathbf{S}_{\mathbf{r}_S}\left[\mathbf{w}_H^T\mathbf{H}_{\mathbf{r}_H}\left[x,\mathbf{z}_{H_0}\right]\right](t),\qquad(A.5)$$

where $n$ and $l$ represent the order of the model.

## A.2. Definition of the inverse MPI model

One of the main features of the MPI model is that there is a strict relationship between the direct and the inverse model.

The inverse PI model is defined as:

$$H^{-1}[y](t) \triangleq \mathbf{w}_H'^T \mathbf{H}_{\mathbf{r}_H'}\left[y, \mathbf{z}_{H_0}'\right](t), \tag{A.6}$$

where inverse weight vector $\mathbf{w}_H'$ can be obtained by the weights and threshould values of the direct model. Similarly, also the superposition operator is invertible, and its inverse is defined as:

$$S^{-1}[y](t) \triangleq \mathbf{w}_S'^T \mathbf{S}_{\mathbf{r}_S'}[y](t), \tag{A.7}$$

whose weights are still obtainable by the parameters of the direct superposition operator.

This leads to the inverse MPI model, which is defined as:

$$\Gamma^{-1}[y](t) \triangleq \mathbf{w}_H'^T \mathbf{H}_{\mathbf{r}_H'}\left[\mathbf{w}_S'^T \mathbf{S}_{\mathbf{r}_S'}[y], \mathbf{z}_{H_0}'\right](t). \tag{A.8}$$

## A.3. Identification of the MPI model

In order to identify an hysteretic characteristic, the first step is to determine the ranges in which input and output will be contained. Based on such information, the threshould values of the PI operator and the threshould values of the inverse superposition operator are determined. In fact, while the former will be uniformly distributed over the range of the input, the latter will be uniformly distributed over the range of the output.

In fact, the identification procedure consists in identifying the weights of the direct PI operator and the weights of the inverse superposition operator, since the identification error can be expressed as:

$$E[x,y](t) \triangleq \begin{pmatrix} \mathbf{w}_H^T & \mathbf{w}_S'^T \end{pmatrix} \begin{pmatrix} \mathbf{H}_{\mathbf{r}_H} \big[ x, \mathbf{z}_{H_0} \big](t) \\ -\mathbf{S}_{\mathbf{r}_S'} [y](t) \end{pmatrix}, \tag{A.9}$$

which makes use of the property of monotonicity of the superposition operator. This definition of the error leads to the quadratic optimization problem

$$\min \left\{ \begin{pmatrix} \mathbf{w}_H^T & \mathbf{w}_S'^T \end{pmatrix} \int_{t_0}^{t_1} \begin{pmatrix} \mathbf{H}_{\mathbf{r}_H} \big[ x, \mathbf{z}_{H_0} \big](t) \\ -\mathbf{S}_{\mathbf{r}_S'} [y](t) \end{pmatrix} \cdot \left( \mathbf{H}_{\mathbf{r}_H} \big[ x, \mathbf{z}_{H_0} \big](t)^T \quad -\mathbf{S}_{\mathbf{r}_S'} [y](t)^T \right) dt \begin{pmatrix} \mathbf{w}_H \\ \mathbf{w}_S' \end{pmatrix} \right\}. \tag{A.10}$$

This minimization problem is overdetermined with one degree of freedom since the elementary operators $H_{r_H}\big|_{r_H=0}$ and $S_{r_S'}\big|_{r_S'=0}$ are both equal to the identity operator $I$. The additional linear equality constraint

$$\left( \left( \|x\|_\infty \cdot \mathbf{i} - \mathbf{r}_H \right)^T \quad \mathbf{o}^T \right) \begin{pmatrix} \mathbf{w}_H \\ \mathbf{w}_S' \end{pmatrix} - \|x\|_\infty = 0 \tag{A.11}$$

with the unity vector $\mathbf{i}^T = \begin{pmatrix} 1 & 1 & \dots & 1 \end{pmatrix}$ deletes this degree of freedom and ensures the unique solvability of the quadratic minimization problem.

After identifying the direct PI operator and the inverse superposition operator, their inverse operators can be obtained by direct relationships [Kuh01]. Thus, the identification leads in one step both to the direct and inverse model.

## A.4. A MPI model with creep

The MPI model can be extended to model also the creep effect of many active materials. Let $x$ be the input, $y_i$ the output, $r_{K_i}$ the threshould value and $a_{K_i}$ the creep eigenvalue, the elementary *creep* operator for a generic instant of time $t$ $y_i(t) = K_{r_{K_i} a_{K_i}} \left[ x, y_{i_{K_0}} \right](t)$ is defined as the unique solution of the nonlinear differential equation

$$\frac{d}{dt} y_i(t) = a_{K_i} \max\left\{ x(t) - y_i(t) - r_{K_i}, \min\left\{ x(t) - y_i(t) + r_{K_i}, 0 \right\} \right\}, \quad (A.12)$$

with the initial condition

$$y_i(t_0) = y_{i_{K_0}}. \tag{A.13}$$

In Figure A.3 the step response of the elementary creep operator over the logarithmic time axis for a number $m$ of creep values $a_{K_j}$ is shown.

In this case the creep eigenvalues are distributed exponentially over the reciprocal of the sampling time $T_s$ according to the following equation:

$$a_{K_j} = \frac{1}{10^{j-1} T_s}; \qquad j = 1\ldots m, \tag{A.14}$$

where $m$ is the order of the creep operator.

The PI creep operator for complex $\log(t)$-type creep effects is the weighted sum of $n+1$ elementary creep operators and it is defined as:

$$K[x](t) \triangleq \mathbf{w}_K^T \mathbf{K}_{r_K a_k} \left[ x, \mathbf{z}_{K_0} \right](t) \cdot \mathbf{i}, \tag{A.15}$$

where $\mathbf{w}_K^T = \left( w_{K_0}, w_{K_1}, \ldots, w_{K_n} \right)$ is the weight vector, $\mathbf{r}_K^T = \left( r_{K_0}, r_{K_1}, \ldots, r_{K_n} \right)$

is the threshould vector with the property that $0 < r_{K_0} < r_{K_1} < \cdots < r_{K_n} < +\infty$, $\mathbf{a}_K^T = \left( a_{K_1}, \ldots, a_{K_m} \right)$ is the vector of creep eigenvalues, $\mathbf{i}^T = \begin{pmatrix} 1 & 1 & \cdots & 1 \end{pmatrix}$ is the identity vector, and the matrix of initial conditions and the matrix of elementary creep operators are defined respectively as:

$$
\mathbf{Z}_{K_0} = \begin{pmatrix}
z_{K_{0_{01}}} & \cdots & z_{K_{0_{0m}}} \\
z_{K_{0_{11}}} & \cdots & z_{K_{0_{1m}}} \\
\vdots & \ddots & \vdots \\
z_{K_{0_{n1}}} & \cdots & z_{K_{0_{nm}}}
\end{pmatrix},
\tag{A.16}
$$

$$
\mathbf{K}_{r_K a_K}\left[ x, \mathbf{z}_{K_0} \right] = \begin{pmatrix}
K_{r_{K_0} a_{K_1}}\left[ x, z_{K_{0_{01}}} \right] & \cdots & K_{r_{K_0} a_{Km}}\left[ x, z_{K_{0_{0m}}} \right] \\
K_{r_{K_1} a_{K_1}}\left[ x, z_{K_{0_{01}}} \right] & \cdots & K_{r_{K_1} a_{Km}}\left[ x, z_{K_{0_{1m}}} \right] \\
\vdots & \ddots & \vdots \\
K_{r_{K_n} a_{K_1}}\left[ x, z_{K_{0_{n1}}} \right] & \cdots & K_{r_{K_n} a_{Km}}\left[ x, z_{K_{0_{nm}}} \right]
\end{pmatrix}.
\tag{A.17}
$$

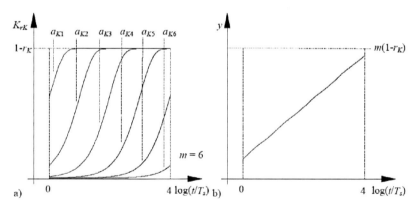

**Figure A.3 – Step response of a) the elementary creep operator and b) a log(t)-type creep operator.**

The complete MPI model with creep is then redefined as:

$$\Gamma[x](t) \triangleq S\big[H[x](t) + K[x](t)\big](t) =$$
$$\mathbf{w}_S^T \mathbf{S}_{\mathbf{r}_S} \Big[\mathbf{w}_H^T \mathbf{H}_{\mathbf{r}_H} \big[x, \mathbf{z}_{H_0}\big] + \mathbf{w}_K^T \mathbf{K}_{\mathbf{r}_K \mathbf{a}_K} \big[x, \mathbf{z}_{K_0}\big] \mathbf{i}\Big](t), \tag{A.18}$$

which still admits its inverse model which is redefined as:

$$\Gamma^{-1}[y](t) \triangleq \mathbf{w}_H^{\prime T} \mathbf{H}_{\mathbf{r}_H'} \Big[\mathbf{w}_S^{\prime T} \mathbf{S}_{\mathbf{r}_S'} \big[y\big] - \mathbf{w}_K^T \mathbf{K}_{\mathbf{r}_K \mathbf{a}_K} \big[x, \mathbf{z}_{K_0}\big] \mathbf{i}, \mathbf{z}_{H_0}'\Big](t). \tag{A.19}$$

The identification procedure shown in section A.3 extends simply to the MPI with creep.

## *A.5. Identification example*

Let us consider the actuation characteristic of a piezoelectric material, which is affected by creep. Moreover, in order to evaluate the quality of the identification, let us define the identification error *err* as:

$$err = \frac{\big\|\Gamma[x](t) - y(t)\big\|_\infty}{\big\|\Gamma[x](t)\big\|_\infty}. \tag{A.20}$$

Figure A.4 shows the normalized applied voltage and the normalized measured elongation. As one can see, the identification voltage is designed in a way to excite the hysteretic characteristic at several amplitudes. Moreover, its step construction enhances the visibility of creep. Indeed, the measured elongation is affected by asymmetric hysteresis and creep. A MPI model with creep of order $n = 8$, $l = 4$ and $m = 1$ has been identified. Figure A.5 shows the identified characteristic and the identification error over time. The identification error is particularly low, since it is within a 1% band which is acceptable for most of the applications.

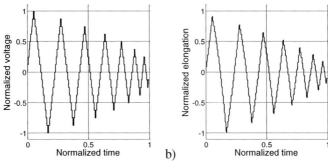

a)          b)

*Figure A.4 – Normalized a) voltage and b) elongation vs. normalized time scale.*

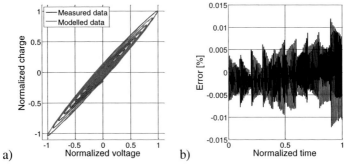

a)          b)

*Figure A.5 – Identification results. A) Normalized voltage and elongation characteristic b) error vs. normalized time scale.*

# B.   Vibration control

Vibration control of flexible structures is an important issue for many engineering applications. In fact, vibration reduction is often desirable for avoiding structural damages, reducing noise of working environments, increasing performance and comfort in many applications, from aerospatial to automotive and so on.

The aim of a vibration controller is mainly to introduce damping at the resonances that characterize a flexible vibrating structure (see Chapter 2). Generally, it is desirable to design multi-mode controllers, which are capable to damp more than one resonance frequency at the same time, while avoiding to introduce undesired behaviour at higher frequencies (spill-over effect) [TS06].

There are three approaches to introduce damping into a vibrating structures: passive, active and semi-active. Passive methods make use of actuators which do not require any kind of external power nor sensors [HMJ01]. Active methods, instead, use actuators capable of generating active forces on the structure which counteract the vibration. A vibration controller receives the sensor information and solve the control law necessary to control the actuator. Finally, semi-active methods join the capabilities of the passive and active methods, by using passive devices which parameters can be tuned according to the need. This is the case of magnetorheological fluid actuators, whose viscosity can change according to the magnetic field passing through the fluid [Pre02], or semi-active shunt damping circuitry, which is a tunable electrical load connected to a piezoelectric material integrated into the vibrating structure which is intended to dissipate the electrical energy which is obtained by electro-mechanical conversion of the piezoelectric material.

Digital control systems allow the implementation of complex control law.

However, due to the sampling frequency, a delay is always introduced in the control loop, which degrades the controller performance. Analogical controllers, instead, can typically have better performance than the digital ones, especially at high frequencies, where the effects of the delay introduced by the digital systems become consistent, due to the absence of anti-aliasing filters, ADCs and DACs and sampling discretization. Nevertheless, it is not always easy to implement complex control algorithms digitally, and the necessary effort is not always spendable.

In this section, two vibration controllers are presented: the Positive Position Feedback (PPF) controller and the Resonant controller.

## B.1. Positive Position Feedback (PPF) controller

The PPF controller has been firstly introduced in 1982 in order to control the vibrations of large flexible space structures [SSA01]. It found then great success in vibration control applications because of the simplicity of tuning and its robustness respect to damping rations and resonant frequencies knowledge [FL08], [HSJ10].

The name of this controller originates from the peculiarity that the sensor information $\xi$ is positively fed back to the controller, which is unusual in control loops, as it is shown in figure B.1.

Let us consider the scalar case of a single mode vibrating structure and a single actuator. The system consists of two equations, respectively the structure and controller differential equations:

$$\ddot{\xi} + 2\delta\omega\dot{\xi} + \omega^2\xi = g\omega_f^2\eta, \qquad (B.1)$$

$$\ddot{\eta} + 2\delta_f\omega_f\dot{\eta} + \omega_f^2\eta = \omega_f^2\xi, \qquad (B.2)$$

where $g$ is a scalar gain greater than 0, $\xi$ and $\eta$ are respectively the modal and filter coordinate, $\omega$ and $\omega_f$ respectively the modal and filter frequencies, $\delta$ and $\delta_f$ respectively the modal and filter damping ratios. The reference signal is then set to zero as usual in vibration control.

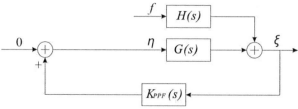

*Figure B.1 – Single-mode PPF control scheme.*

The stability of such control system is guaranteed for $0 < g < 1$ [FC90]. The transfer function of a single-mode PPF controller can then be expressed in the Laplace domain as:

$$K_{PPF}(s) = \frac{g\omega_f^2}{s^2 + 2\delta_f \omega_f s + \omega_f^2},$$ (B.3)

which is a second order low pass filter. The low pass filtering action is very important, since any high frequency disturbance in sensing path would be cut out of the control law. Thus spillover reduction is obtained. An example of frequency response of such controller is shown in Figure B.2.

Let us consider now the case of a vibrating structures with $N$ modes. Since a PPF controller adds damping only to one mode, it is easily extendable for multi-mode vibration damping. In fact, a multi-mode PPF controller $K_{mPPF}(s)$ can be obtained by summing the output of $N$ single-mode PPF controllers, each one tuned for a particular resonant frequency, as shown in the following equation:

$$K_{mPPF}(s) = \sum_{k=1}^{N} K_{PPFk}(s) = \sum_{k=1}^{N} \frac{g_k \omega_{fk}^2}{s^2 + 2\delta_{fk} \omega_{fk} s + \omega_{fk}^2}.$$ (B.4)

A schematic of multi-mode PPF control of a vibrating structure is drawn in Figure B.3.

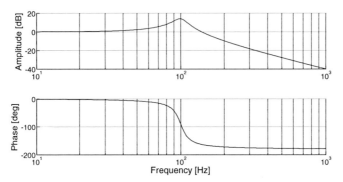

Figure B.2 – Frequency response of a PPF controller tuned at 100 Hz with δ = 0.1 as damping ratio.

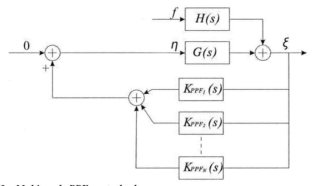

Figure B.3 – Multi-mode PPF control scheme.

A stability criterion for the multi-mode approach is not as simple as for the single-mode case, and it will not be treated in this work. Nevertheless, in the case the resonant frequencies to be damped are far enough from each other so that each PPF controller does not affect the other ones, the stability criterion can be handled per each controller as in the single-mode case.

*Optimal tuning for a PPF controller*

A single-mode PPF controller can be tuned by adjusting the gain $g$, the damping ratio $\delta_f$ and the frequency $\omega_f$. Generally, the controller frequency is set equal to 1.1 up to 1.4 times the structural resonance, while the damping ratio is usually found in the range 0.01 ... 0.5. Thus, the controller gain $g$ is found by trial and error. Nevertheless, McEver proposed a method in 1999 [Mce99] which is based on the knowledge of the structural pole/zero pairs. Let us consider a structural transfer function with two zeros and two poles, defined as follows:

$$G(s) = \frac{s^2 + \omega_z^2}{s^2 + \omega_p^2}, \tag{B.5}$$

where $\omega_z = \alpha\omega_p$ and $\alpha > 1$. The closed loop transfer function $G_{CL}(s)$ is then:

$$G_{CL}(s) = \frac{G(s)}{1 - G(s)K_{PPF}(s)}. \tag{B.6}$$

The optimal parameters of the controller are chosen by comparing the coefficients of the denominator of the closed loop transfer function $G_{CL}(s)$ and a desired denominator. Consequently, one obtains the following relations:

$$\omega_f = \omega_p \sqrt{\frac{1}{1 - g\alpha^2}}, \tag{B.7}$$

$$\delta_f = \frac{\omega_p \sqrt{\frac{1-g}{1-g\alpha^2} - 1}}{\omega_f}. \tag{B.8}$$

The gain is then chosen according to the gain margin of the system. In fact the open loop transfer function $G_{OL}(s)$ is expressed as:

$$G_{OL}(s) = K_{PPF}(s)G(s) = \frac{g\omega_f^2}{s^2 + 2\delta_f \omega_f s + \omega_f^2} \frac{s^2 + \omega_z^2}{s^2 + \omega_p^2}, \tag{B.9}$$

whose phase can be $-180°$ only for $s = 0$. The gain margin $GM$ is then related to the PPF controller gain $g$ according to the following equation:

$$GM = \frac{1}{g\alpha^2}, \tag{B.10}$$

which has to be greater than one to guarantee the stability of the system.

Let us consider the case of a vibrating structure whose transfer function is of the type shown in equation (B.5) (resonance frequency at 40 Hz and antiresonance frequency at 45 Hz), where some small damping is introduced (0.004), and a single-mode PPF controller is tuned according to the above relations and the $GM$ is, for example, set to 3. The frequency response of the system with and without control is finally shown in Figure B.4.

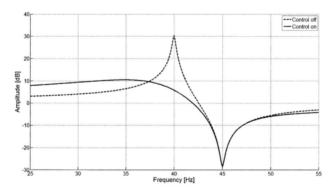

*Figure B.4 – Frequency response amplitude of the vibrating structure with and without PPF control.*

## B.2. Resonant controller

PPF controllers have the advantage of not introducing spill-over effects. Nevertheless, each single-mode PPF controller affects the other ones and the stability of the overall system can be difficult to ensure. Tuning a multi-mode PPF controller is in fact a complex procedure, even though the tuning procedure for a single-mode PPF controller is quite easy. Resonant controllers (initially proposed in [RPS99]) are optimized for working in multi-mode configuration for the control of flexible structures. In fact, its peculiarity is to apply a high gain at the resonant frequency to roll off then quickly far from resonance.

A single-mode resonant controller $K_{RC}(s)$ is defined as:

$$K_{RC}(s) = g \frac{s^2 + 2\delta_f \omega_f s}{s^2 + 2\delta_f \omega_f s + \omega_f^2}, \qquad (B.11)$$

and its typical frequency response is shown in Figure B.5.

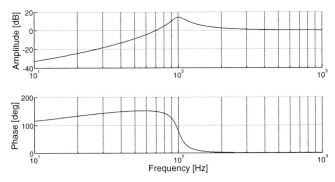

*Figure B.5 – Frequency response of a resonant controller tuned at 100 Hz with $\delta = 0.1$ as damping ratio.*

As in the case of PPF controllers, also multi-mode resonant controllers are obtained by summing the output of many single-mode resonant controllers. Thus, considering $N$ modes to be damped, the multi-mode resonant controller $K_{mRC}(s)$ is obtained as:

$$K_{mRC}(s) = \sum_{i=1}^{N} K_{RC}(s) = \sum_{i=1}^{N} g_i \frac{s^2 + 2\delta_{fi}\omega_{fi}s}{s^2 + 2\delta_{fi}\omega_{fi}s + \omega_{fi}^2}. \qquad (B.12)$$

A schematic of multi-mode resonant control of a vibrating structure is drawn in Figure B.6.

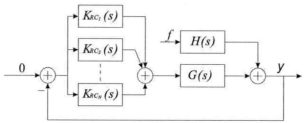

**Figure B.6 – Multi-mode resonant control scheme.**

The frequency response of the resonant controller is specular to the one of a PPF controller. In fact, in this case the low frequencies are filtered, behaving as a high-pass filter. Also in this case, three parameters need to be tuned: the controller angular frequency $\omega_f$, which are set equal to the structural resonant frequency to be damped, the damping ratio $\delta_f$ and the gain $g$, which are instead found empirically. Generally a low value of the damping ratio leads to an amplification of the vibration at frequency far from the resonant one. This value is then usually chosen between 0.01 and 0.5. Moreover, a high gain corresponds to higher attenuations at the resonant frequency.

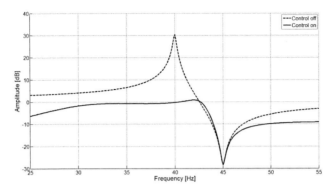

*Figure B.7 – Frequency response amplitude of the vibrating structure with and without resonant control.*

Considering the fact that each single-mode resonant controller gives a static contribution at high frequencies, in the multi-mode configuration such static contributions are summed. Therefore, gains for high frequency controllers are usually smaller than the ones tuned at low frequencies.

Finally, in Figure B.7 the amplitude of the frequency response of a resonant controlled vibrating structure is shown, by using the same vibrating structure of the example in section B.1.